From the Paris Agreement to a Low-Carbon Bretton Woods

Michele Stua

From the Paris Agreement to a Low-Carbon Bretton Woods

Rationale for the Establishment of a Mitigation Alliance

Springer

Michele Stua
C+3C Sistemi e Strategie srl
Venezia, Italy

The views expressed in this book are attributable only to the author in a personal capacity, and not to any institution with which he is associated, or to the funders or supporters of the paper.

ISBN 978-3-319-54698-8 ISBN 978-3-319-54699-5 (eBook)
DOI 10.1007/978-3-319-54699-5

Library of Congress Control Number: 2017933478

Printed on acid-free paper

This Springer imprint is published by Springer Nature
The registered company is Springer International Publishing AG
The registered company address is: Gewerbestrasse 11, 6330 Cham, Switzerland

A Federico

With so much at stake, and with such enormous difficulties of inducing countries to surrender their sovereignty to the global good, surely it pays to look back carefully at history.

(Maurice David Levi)

Foreword

Twenty short years ago, an American delegation, inspired by the adoption of the Clean Air Act of 1990 that established an emission trading system for sulphur emissions in the north-eastern states of the USA, had the Clean Development Mechanism (CDM) inserted in the Kyoto Protocol. Since then, the idea of trading in emission rights has been prominent in all talks about reduction of GHG emissions – this in spite of the fact that trading concededly has never had, and was never intended to have, any emissions reduction outcome (some would even argue to the contrary) and this also despite the recent demise of the CDM, suffering many ills, but in the end killed by starvation (in the sense that markets had lost their appetite).

Hardly by design, the Paris Agreement reinvigorated if not the CDM, then the idea that trading is part of the solution. Although Article 6 is careful to avoid any reference, the CDM looms large. Although wording on 'net reduction' is included, possibly to comfort those that are convinced of the counterproductivity of trading, it is nevertheless a basis for a new trading regime at risk of repeating history. 'Those who cannot remember the past are condemned to repeat it'. Article 6, however, is not a likely result of a failing memory. Hopefully it is not the result of a selective one either. The CDM was immensely successful in some areas – but fatally flawed in others.

It is worth considering what objective trading serves. The CDM was burdened with a multitude of objectives, not least technology transfer and cost-efficiency, but not emissions reduction as any emissions reduction, real or calculated, would be offset in the buyer's account. The post-rationale has been that Kyoto's global target would have been lower without CDM included. A similar rationale may eventually emerge for the Article 6 mechanisms even if the Nationally Determined Contributions (NDCs) were determined long before Article 6 was predicted. Surely there will be claims that the future updating of NDC will show higher ambition than would have been the case in the absence of Article 6. Such battles are fruitless. But they do relate to the most important part of carbon trading, the part that creates the demand for offsets in the first place, the emissions constraint imposed through the cap in a cap-and-trade system.

In this regard, the message of this book is important, not so much for its embrace of Article 6, but for articulating in which context Article 6 could, and probably should, work. I always claimed that if anyone wanted to see what CDM would look like at a level playing field, look at JI, not that the JI was flawless, but at least that the trading was a zero sum game. Allowances sold were intended to reduce the seller's emission budget. Only in such a system, and only if the budget restriction is tangible, can the value of trading be determined. The author addresses this point excellently, while Article 6 is silent. Among all the possible learnings from CDM, this is probably the most important: trading must succeed among equal partners, possibly in an Alliance as proposed in this book, among which emission budgets have been determined transparently. Ultimately, it is the emission budget that determines if emissions are reduced, not the trading. If a trading option helps reaching agreement on the budget, it is of course helpful. Only it must be asked: What is the expected benefit that lends such importance to trading?

There have been many attempts to document the benefits of CDM. But beyond its indisputable achievement in bringing the emissions reduction agenda to all corners of the world, there is no uniform conclusion. If a Mitigation Alliance is established among relatively equal partners, then the argument of technology transfer is likely to become mute. Rather the argument is likely to relate to a cost-efficient emissions reduction response. There is, however, only limited merit in the cost-efficiency argument. Throughout the CDM heydays, low-cost industrial gas projects were developed in parallel to high-cost hydro and wind energy projects with no particular price differences among resulting CERs. Although price formation theories may be employed to explain the phenomenon, it is hard to conclude that the markets were in clear pursuit of cost-efficiency.

This volume points out that the Paris Agreement was the surprise compromise that brought multilateralism back on track but rightfully questions if that is a good or a bad thing. It is not difficult to draw up a scenario, in which failure would have prompted a de facto 'Mitigation Alliance', in face of imminent disaster, to take action. With the Paris Agreement, unfortunately, in spite of the rapid depletion of the global emission budget, there is still no trace of a global burden sharing agreement. Although the Paris Agreement with 190+ countries involved is a major diplomatic accomplishment, the principles upon which such burden sharing should succeed are never to be agreed as long as there are rents to be earned from disagreement – and as long as the cost of disagreement is poorly understood and to be paid by the next generation.

Therefore, in the scenario developed here, the Mitigation Alliance still plays a central role in bringing about substantial emissions reduction. Although the Alliance is no longer a stand-alone structure and operates under the umbrella of the Paris Agreement, it does not depend on successful completion of UN-led negotiations that are to continuously ramp up ambition in updated NDCs in order to close the gap between current emissions and the 1.5–2-degree boundary. It is to be assumed that the Mitigation Alliance will constantly outperform the average NDC ambition level, which may either serve as inspiration for those outside the Alliance or as relief for those that consider themselves less obligated.

Possibly, the Achilles heel of such alliance building is its attraction in practice. As Stua suggests, it depends on a strong governance system to which members subject themselves, including an objective burden sharing. Subjecting oneself to these conditions, the Mitigation Alliance concedes to its members' usage of Article 6. But in face of collective shortage, even individual shortage, who will feel impelled to trade? If the Alliance is to become net-zero carbon over time, possibly a very short time, who will give up a part of its budget to serve the needs of a less fortunate Alliance member? Musketeer oaths are regrettably out of fashion. Trade propensities in CDM were given as deals were one-sided in terms of hollowing or expanding emissions budgets, as the seller had an endless budget to sell from. Equally, in JI, the common seller had excessive budgets of hot air, the selling of which would not leave the seller constrained in any foreseeable future. But when Alliance members are to trade over the last tons of certified mitigation outcomes and everybody is in dire straits, you may have the money to buy, but nobody is selling.

Stua experiments with ways out of this dilemma, including the opportunity to transform the certified mitigation outcome into a representative currency within the Alliance. Attributing a value to environmental, economic, financial, social and political co-benefits may be a way to overcome trade reluctance. This effectively repackages earlier thoughts of using trade policy as a lever to create tangible consequences of unabated emissions. But instead of a punishment for inaction is here a reward for action. The challenge, of course, is if such virtual 'dollar-shops'[1] can be established, if the goods in the shop are in clear demand and if it can be ensured that either the exchange rate to monetary currency significantly favours those in possession of the outcomes or if there is no exchange possible at all.

The Bretton Woods idea presented in this volume addresses this challenge excellently – but it would be a pity to give it all away in a foreword. Read Chap. 8 for yourself. A commodity-based system – not like the gold standard from current emissions trading – is one key element in which the real value comes from the finite supply of CMO: the difference between current and net-zero emissions. No doubt that is ultimately where we need to go – or at least a very large majority of us – meaning that to be effective a Mitigation Alliance must include all major emitters. That is the foremost challenge to any system. No matter the sugar-coating, many might find that the pill is still a bitter one. For that, Stua's reward system could well show the way forward.

For sure the proposed Bretton Woods model could lend more (and needed) credibility to an international agreement like the Paris Agreement, although those that have grown sceptical over the past 25 years of climate debate and negotiations would probably point to the fact that compared to gold, even if we all agree that the CMO is the basic and finite commodity, it remains intangible. But then read the final chapters. Although punishment specifically is not the approach, it could be considered de facto 'punishment' not to be part of the Alliance.

[1]Dollar-shops is used for shops, in currency-constrained countries, that trade in imported goods but only accept foreign currency, typically USD, as payment.

From that perspective, it is not necessarily a given that the forum for creating the Alliance is a UN body like the UNFCCC. Many had already buried the Convention as a useful platform, when Paris produced its surprise compromise. With Paris in place, however, it is wise to embrace it, as Stua does by adopting the CMO, while not necessarily considering it sufficient. The near future will probably tell. Under all circumstances, even if our climate change negotiators have precious few successes to show for themselves, we must remain positive that it can still be done.

UNEP DTU Partnership
Technical University of Denmark
Marmorvej 51, København, Denmark

Søren E. Lütken

Acknowledgements

The task of correctly crediting all those mentors, colleagues and friends that have supported me on this pathway over the last 10 years is even more ambitious than that of expressing the idea contained in this book. Nevertheless, I feel the need to specifically mention a minimum and indeed incomplete number of persons that have played a significant role in making this all possible. My first, deep acknowledgement goes to those collaborators who worked with me on the actual definition of this book: Axel Michaelowa, Researcher at the University of Zurich and Managing Director Perspectives Climate Change, who kindly wrote the introduction of this book and, more importantly, represented a constant reference for the development of this work over the years; Peter Zaman, Partner at Reed Smith LLC, who was a continuous help and proved a precious source of advice in the elaboration of the most delicate sections of the book; Michael Coulon, Lecturer in Finance at the University of Sussex, whose sincere interest in the idea since its very early stages represented a fundamental contribution to its realisation and whose advice and vision transformed the formula into a functional policy tool; and Colin Nolden, whose devoted passion, significant skills and capability of understanding have provided an indispensable and priceless contribution to the development and writing of the entire book. A special thank goes also to Søren Lütken, Senior Economist at UNEP DTU, who kindly accepted to write the foreword, demonstrating sincere interest and passion for the idea.

Two special friends plotted a successful plan to convince me writing this book in the first place, hence deserving my deepest gratitude: Alessandro Costa, Co-Founder and Associate at Sostenibilità e Energia, C + 3C Sistemi e Strategie Consultancies, and Nicola Tollin, Senior Analyst at RESURBE International Program on Urban Resilience.

This proposal would have never seen the light without the support of two special institutions and the sincere friendships found therein. In an attempt to express my gratitude to these institutions, I would like to thank Nicola Bellini, Marco Frey, Andrea Piccaluga, Fabio Iraldo and the whole Sant'Anna School of Advanced Studies in Pisa, Italy, where this all started. My sincere appreciation goes also to Gordon MacKerron, Jenny Lieu and Frank Brouwer, who kindly supported and encouraged me over the years, and all those close friends I found at the University of Sussex.

My gratitude goes to two special angels, Máire Mulloy and Christie Smith, who managed to transform my raw and somewhat convoluted writing style into an understandable and effective English.

This proposal is part and parcel as well as further development of an idea originally put forward by Alfredo Sirkis, Founder and Director of the think tank Centro Brasil no Clima. As Congressman in Brazil, he played an important role in drafting several of the foundational ideas in this book, and he has provided immense support for the evolution of these ideas in general and this book project in particular. Everton Lucero, Vice-Minister for Climate Change and Environment Quality at the Brazilian Ministry of Environment, engaged in the discussions about the opportunity to implement the proposal discussed in this book and promoted several of its elements and more importantly played a key role in the Brazilian Submission that led to the inclusion of Article 6 within the Paris Agreement. Maurice David Levi, whose visionary paper of 1991 represents a seminal reference for this book, requires a special tribute too. Despite the limited opportunities to meet and exchange opinions, Christophe de Gouvello, Senior Energy and Climate Change Specialist at World Bank, has been a precious point of reference in recent years. I will never forget that two innovative minds, as well as close and special friends, helped me to confront several issues that impacted the correct evolution of the proposal contained in this book: Diego Isabel La Moneda, Co-Founder and Director of the Global Hub for the Common Good, and Julián Piñeres Ramirez. Finally, I wish to thank with all my heart the three persons without whom none of what made this book possible would have ever happened: my parents Graziella and Franco and the reference of my life Serena, y.s.k.

Collaborator and Co-Author

Dr. Colin Nolden
Colin is an independent researcher focusing on energy policy, climate policy and sustainability transitions. He has been recently working on Climate-KIC's Transition Cities Project. Previous appointments include working as a researcher at Science Policy Research Unit (SPRU), University of Sussex; Centre for Rural Policy Research, University of Exeter; and Geography College of Life and Environmental Sciences, University of Exeter; working as an energy consultant for German energy agencies and the UK government; as well as stints working as a steelworker and a mussel farmer. His academic background spans geography, history, economics and sustainable development. Since 2014 he has engaged with the concept of carbon markets.

Collaborators

Peter Zaman (https://www.reedsmith.com/Peter_Zaman/)
Peter is a transactional lawyer who practices in commodity derivatives, derivatives and structured products and carbon finance. With more than 13 years of experience in the commodities markets, Peter has used his experience to build a market-leading reputation as consistently recognised by *Chambers and Partners UK and Global* (2009–2015), in the London Super Lawyers List for Energy and Natural Resources (2014–2015), as a leading individual for commodities (general) by *Legal 500 UK* (2014–2015), and by PLC's Which Lawyer (2011) and is listed in *Who's Who Legal: Environment* 2015. He acts for commodity merchants, miners, utilities, investment banks, multilateral development institutions, corporates and fund managers. Peter advises on all aspects of over-the-counter, exchange-traded and structured energy derivatives and commodities, including collateral structures in commodity transactions and standby letters of credit. Peter's commodity product suite includes LNG, base metals, gas, electricity and coal. Peter is also recognised as a market leader in climate change, carbon trading and finance. His carbon specialisation extends to all aspects of carbon finance (Green Climate Fund, CDM, JI, Green Investment Schemes etc.), green bonds, emissions trading and sustainable commodity funds. He is a recognised and established authority on legal issues relating to emissions trading and has been involved in the area since the inception of the EU Emissions Trading Scheme in 2005. Peter's financial derivatives and structured products experience arises from his involvement in advising lenders, borrowers, issuers and arrangers in the context of the derivative aspects of project finance, real estate finance and capital markets products. He has been an active participant at ISDA and also advises on GMRAs, GMLSAs, prime brokerage arrangements and other structured derivative products.

Dr. Michael Coulon (http://www.sussex.ac.uk/profiles/329300)
Michael is a senior lecturer in finance in the Department of Business and Management, University of Sussex. Prior to joining the University of Sussex in July 2013, Michael worked as a postdoctoral research associate and associate research scholar at Princeton University (2009–2013), in the Department of Operations Research and Financial Engineering (ORFE). His postdoc was supported by the National Science Foundation (NSF) through Princeton's research training group (RTG) in stochastic analysis and applications. Michael completed a PhD (DPhil) in mathematics in 2009 at the University of Oxford. He also holds a master in finance degree from Princeton University and a BSc in mathematics from Imperial College, London. He has also worked for short periods in the financial industry, at JPMorgan and Goldman Sachs. Michael's research activities centre on quantitative finance and its application to energy and other commodity markets. He is an associate editor at the *Journal of Commodity Markets*, launched in 2015. He has developed various commodity price models and derivative pricing techniques, especially for wholesale electricity markets, as well as environmental markets such as carbon allowances and green certificates. He has designed new and innovative teaching material on such topics, regularly presents his research at leading international conferences in this field and has published in top international journals. He has also collaborated with a variety of energy companies and other industry partners to apply mathematical techniques to practical risk management and valuation problems.

Dr. Axel Michaelowa (http://www.perspectives.cc/index.php?id=672)
Axel is head of the group on international climate policy at the chair 'Political Economy and Development', University of Zurich and Center for Comparative and International Studies (CIS). Besides his research work, he is senior founding partner of the consultancy 'Perspectives'. Axel is a lead author of the chapter on international agreements in the 5th Assessment Report of the Intergovernmental Panel on Climate Change and wrote on mitigation policies in the 4th Assessment Report of the Intergovernmental Panel on Climate Change published in 2007. He is member of the editorial boards of the journals *Climate Policy*, *Climate and Development* and *Greenhouse Gas Measurement and Management* as well as of the board of the research network 'Climate Strategies'. For the last two decades, Axel has been working on climate policy issues and published over 100 articles, studies and book contributions, mainly on the Kyoto Mechanisms and especially the Clean Development Mechanism (CDM). Axel is a member of the CDM Executive Board's Registration and Issuance Team and has contributed to the development of eight approved CDM baseline methodologies. As one of the leading European experts on CDM institution building in developing countries, Axel supported several governments in Asia and North Africa/Middle East.

Contents

Editor and Main Author

Dr. Michele Stua Born in Udine, Italy, on 20 April 1973, Michele Stua is a researcher and analyst focusing on Climate Policy. Currently affiliated to C+3C Sistemi e Strategie srl, Michele Stua has worked for Sant'Anna School for Advanced Studies in Pisa, where he obtained his Ph.D. in 2011, developing an innovative research on the application of the Clean Development Mechanism (CDM) in China, and SPRU–Science Policy Research Unit, University of Sussex. His major efforts concerned the study of carbon mechanisms aimed at facilitating the trade of mitigation-related assets worldwide. His findings led to the definition of policy proposals introduced in different international contexts.

He directly collaborated with Brazilian authorities to design the proposals contained in the Brazil submission to the United Nations Framework Convention on Climate Change (UNFCCC) of November 2014, which eventually led to the inclusion of the Article 6 in the new International Agreement on Climate Change signed in Paris on December 2015.

Chapter 1
Introduction

Axel Michaelowa

Abstract The Paris Agreement has provided a new long term basis for international climate policy and erased the Copenhagen failure. However, its success depends on the willingness of each participating government to actually mitigate emissions. A transformational club of countries agreeing on common, stringent mitigation targets that achieve the net zero emissions philosophy of the Agreement could play a trust-building role in the early phases of the Paris regime. By using Art. 6 of the Paris Agreement, this 'Mitigation Alliance' would engage in direct or indirect linking of carbon markets and thus generate a benefit for their members. Independent certification of mitigation achieved should be built on the experiences of the Clean Development Mechanism. From the initially existing different carbon markets, a convergence to a system of linked emission trading schemes with a single international and domestic offset scheme for sectors not covered by trading should be aimed for. This will be governed by a set of institutions that are modelled on successful monetary and trade policy institutions. The next years will give us an opportunity to set the rules of the Paris Mechanisms in a manner that supports emergence of a 'Mitigation Alliance'.

Keywords Paris Agreement (PA) • Transformational clubs • 'Mitigation Alliance' • Independent certification • Emissions trading • Offsets • Governance

1.1 Why Can Market Mechanisms Enhance Mitigation Ambition?

Market mechanisms have been attacked over many years by environmentalists to be a loophole that reduces mitigation ambition (e.g. Böhm et al. 2015). We are of the opinion that well-designed market mechanisms can actually enhance mitigation ambition, particularly as the international climate policy regime develops towards a 'pledge and review' system. This book develops a proposal how under the climate

M. Stua, *From the Paris Agreement to a Low-Carbon Bretton Woods*,
DOI 10.1007/978-3-319-54699-5_1

policy regime developed by the Paris Agreement voluntary action by a group of governments can be built on a market mechanism approach. Each of the sub-sections in this introduction describes one chapter of the book and its principal approach.

1.2 The Paris Agreement

The Paris Agreement (UNFCCC 2015) enabled to end a period in which international climate policy had stalled. It approaches greenhouse gas emission reductions from a bottom up perspective, where all countries propose mitigation targets through their Nationally Determined Contributions (NDCs). This approach enabled to exit the impasse that climate negotiations had been in since the Copenhagen conference and overcomes the "iron curtain" between industrialized and developing countries frozen since 1990 (Bodansky 2016). However, it risks that mitigation policies will be taken hostage due to short-term economic and political pressures (Clemençon 2016; van Asselt 2016). Already now, the UNFCCC has stressed that the current NDCs are still very far from the stringency required to fulfil the targets of the Paris Agreement (UNFCCC 2016). Thus, despite its strong ambition embodied in the target to reach a balance between emissions and sinks in the second half of the century, the Paris Agreement is only as strong as the sum of all countries' action. This also applies to the flexibility mechanisms built in its Article 6. They can only become relevant if a sufficient number of governments is willing to use these mechanisms either on the selling or the buying side.

1.3 Climate Clubs and Where They Can Make a Difference

A key feature of delivering ambition under the Paris Agreement could be through coalitions of countries that want to achieve a higher mitigation ambition than others (Sabel and Victor 2015). These coalitions have been labelled 'transformational clubs' (Weischer et al. 2012). They can certify mitigation achieved within the club, and on that basis engage in a common market-mechanism based approach to mitigation, e.g. through linking of emissions trading systems (Keohane et al. 2015). Also, indirect linking through an offset mechanism would be possible whose credits are accepted by all club members. This would provide a "carrot" for participation. At the same time, club member governments might want to ensure that their companies do not duffer a competitive disadvantage by non-club members, for example through carbon border adjustments. However, such an approach may not be consistent with rules of the World Trade Organization. In the absence of that possibility, club members would need to agree on non-compliance penalties for members. Such non-compliance regime could be built around Art. 6 of the Paris Agreement.

1.4 A 'Mitigation Alliance' as Special Form of a Climate Club

Art. 6 can serve as the basis for a transformational club due to its principles that stress the voluntary nature of collaboration among a subset of governments. This 'Mitigation Alliance' could decide on highly ambitious emission targets for its members that would then be denominated in terms of common emission units. In order to make the club viable in the long term, a joint mitigation target could be designed for the entire club and then allocated to the member states according to previously defined burden sharing criteria. A cooperative approach under Art. 6.2 could be used to generate Internationally Transferable Mitigation Outcomes (ITMOs). This approach would fulfil certification functions. In order to provide incentive for joining the club, club members would not allow use of offset credits coming from outside the club nor acquire any mitigation technologies from non-members. This can of course only be implemented if the key providers of mitigation technologies are club members.

1.5 How to Define the Common Mitigation Target

Given the Paris Agreement's overall long-term target to achieve a balance of emissions and sinks in the second half of the century, national emission targets of the Mitigation Alliance should also reflect this target. This could be achieved by defining a global mitigation pathway consistent with the long-term target and then allocating mitigation targets to the member states. The target would be defined in mitigation contributions over time - not emissions allowances - and would take into account carbon embodied in imports of goods and services. It would be defined according to the concept of "graduation and deepening" (Michaelowa et al. 2005) meaning that the stringency of the target would be defined according to tiers of club members. An adjustment term covering emissions growth of club members that are not subject to any target or that have overshot their target would be applied for each future period and thus ensure that the overall mitigation ambition remains the same.

1.6 Learning from the Clean Development Mechanism

One of the key successes of the Clean Development Mechanism (CDM) under the Kyoto Protocol was the methodological backbone of the mechanism as well as its system of independent verification and certification of mitigation results by accredited third parties. Thus, all transactions under Art. 6 within the Mitigation Alliance should undergo third party verification and certification. This should be the case both for project-based and higher aggregation level based mitigation. Obviously, for the latter new methodological approaches are required, as determination of business-as-usual and additionality are more challenging for highly aggregated mitigation on the level of policy instruments or even entire sectors.

1.7 Kinds of Transactions

Whether there will be full integration of carbon markets under a 'Mitigation Alliance' or whether various forms of interaction coexist remains to be seen. Classically, emissions trading systems (ETS) are likely to form the backbone, complemented by project-based offset systems. The latter can be limited to activities in the specific jurisdiction or have an international scope. To reach the ambition embodied in the national emission targets, ETS with strict allocation regimes would have to provide demand for offset systems harnessing mitigation from sectors not appropriate for ETS. Over time, the existing market mechanisms in the different member states would converge into a system of a single, multi-country ETS with a single international and national offset scheme. Whether only monetary payment is done or whether non-monetary forms of remuneration can also be applied, remains to be seen. Transactions can be taxed in kind or in monetary form, building on precedents under the CDM.

1.8 Governing the Market

Robust accounting and clear specification of members' national emission targets that add up to the joint target of the 'Mitigation Alliance' is a crucial precondition for carbon markets. Thus, all members are to have registries of 'Certified Mitigation Outcomes' (CMOs) functioning according to the same level of robustness. Double counting is to be avoided through tracking of CMOs. Governance would be done in a kind of 'Bretton Woods' system (Levi 1991), with four institutions fulfilling different roles. A 'General Assembly' functioning under the 'one country-one vote' principle would be the highest regulatory authority. Demand for CMOs would be generated by a 'Target Fund' modelled on the International Monetary Fund, which would calculate target allocation and track performance of member states as well as enforce compliance. All governments would have to hold sufficient reserves of CMOs to ensure that their mitigation target is achieved. How far the latter can be achieved would be an issue to elaborate through further research. Certification of offsets as well as collection of a CMO transaction tax would be done by a 'CMO Bank'. A 'CMO Exchange Board' would oversee transactions and function according to the model of the World Trade Organization. Transactions could be done on an exchange, through brokers or direct bilateral trade between CMO generators and final users.

1.9 How Carbon Prices Are to Be Generated

In contrast to pricing of allowances, pricing of CMOs provides a revenue to mitigation service providers. Greenhouse gas mitigation becomes a valuable commodity. Within the 'Mitigation Alliance' a single price for CMOs will evolve, doing away with the hodgepodge of pricing in today's fragmented carbon market world. CMOs may even evolve into a kind of generic reserve currency of the entire club.

1.10 Co-benefits of Mitigation

Co-benefits of the mitigation achieved by the 'Mitigation Alliance' are manifold, including increasing value added of economic activities. Creating jobs reducing local air, water and soil pollutants, mobilizing civil society participation and driving technological innovation and technology transfer. These benefits are enhanced through the high degree of liquidity and the long-term perspective of the CMO market.

1.11 Mobilizing a Transformational Club

The 2000s have shown through the Clean Development Mechanism, an attractive mitigation commodity could be created that mobilized action around the world. A 'Mitigation Alliance' could enlarge this success by several orders of magnitude. Only then, the ambition of the Paris Agreement would be reflected by government action, and private sector incentives. The next years will give us the opportunities to design the rules of Art. 6 in a way that gives a boost to attempts to set up one or several transformational clubs. Let us grasp this opportunity!

References

Bodansky, D. (2016). The legal character of the Paris agreement. *Review of European Community and International Environmental Law, 25*(2), 142–150.

Böhm, S., Brei, V., & Dabhi, S. (2015). EDF Energy's green CSR claims examined: The follies of global carbon commodity chains. *Global Networks, 15*, 87–107.

Clemençon, R. (2016). The two sides of the Paris climate agreement: Dismal failure or historic breakthrough? *Journal of Environment and Development, 25*(1), 3–24.

Keohane, N., Petsonk, A., & Hanafi, A. (2015). Toward a club of carbon markets. *Climatic Change*. doi:10.1007/s10584-015-1506-z.

Levi, M. D. (1991). Bretton woods: Blueprint for a greenhouse gas agreement. *Ecological Economics, 4*, 253–267.

Michaelowa, A., Butzengeiger, S., & Jung, M. (2005). Graduation and deepening: An ambitious post-2012 climate policy scenario. *International Environmental Agreements, 5*, 25–46.

Sabel, C. F., & Victor, D. G. (2015). Governing global problems under uncertainty: Making bottom-up climate policy work. *Climatic Change*. doi:10.1007/s10584-015-1507-y.

UNFCCC. (2015). *Adoption of the Paris agreement*. Bonn: Climate Change Secretariat.

UNFCCC. (2016). *Aggregate effect of the intended nationally determined contributions: An update*. Bonn: Climate Change Secretariat.

Van Asselt, H. (2016). International climate change law in a bottom-up world. *Questions of International Law, 26*, 5–15.

Weischer, L., Morgan, J., & Patel, M. (2012). Climate clubs: Can small groups of countries make a big difference in addressing climate change? *Review of European Community and International Environmental Law, 21*(3), 177–192.

Part I
Foundation of the Mitigation Alliance

Chapter 2
The Road to Paris

Colin Nolden and Michele Stua

Abstract To many extents, the Paris Agreement on Climate Change represents the culmination of a 25-year diplomatic debate involving a multitude of stakeholders, including policy makers, scholars, and a plethora of different interest groups. At the same time, the Paris Agreement embodies the foundation of a global transformation that can lead to the establishment of a low-carbon society and economy worldwide by the end of this century. Understanding how such utopic ambition has been codified and ratified in an international treaty and whether the treaty can trigger low-carbon transformation processes requires first of all an understanding of its history. By providing a summary of the key events that characterised the 25-year pathway that led to the Paris Agreement, this chapter offers the information required to understand the Agreement's ambition. The description of the negotiation process since 1992 sheds light on the dramatic shifts of power that shaped its evolution. The Chapter concludes with a detailed description of the Paris Agreement, its content, its achievements, as well as its most significant unsolved issues.

Keywords Climate change negotiations • UNFCCC • Kyoto Protocol (KP) • Copenhagen • Paris Agreement (PA)

2.1 Introduction

Signed by almost 200 parties in Paris on 12 December 2015, the Paris Agreement (PA) represents the culmination of a complex compromise between scientific findings and policy imperatives, as well as one of the most significant international treaties of the last 100 years. Yet at the same time this culmination represents an important new juncture along the way towards a more climate-friendly society. This book describes and defends one of the possible pathways for international policy to follow in the aftermath of the PA. A clear grasp of the processes that led to the PA is vital to understanding the significance of this new pathway. Chapter 2 is therefore devoted to the multi-decade 'road to Paris' in its various aspects and features.

M. Stua, *From the Paris Agreement to a Low-Carbon Bretton Woods*,
DOI 10.1007/978-3-319-54699-5_2

Human induced climate change was first theorised scientifically in the first half of the eighteenth century, most notably by Joseph Fourier (Taylor 2011). In 1979, over 150 years after Fourier made the connection, the First World Climate Conference in Geneva, Switzerland, recognised human induced climate change as a serious problem (Zillman 2009), although a definite warming trend was not evident at that stage. It took another decade before the Intergovernmental Panel on Climate Change (IPCC), now the leading global climate science organisation, was founded in 1988 (IPCC 2016). Climate change finally entered the political realm with the 1992 Earth Summit in Rio de Janeiro, Brazil (United Nations 1992), where the United Nations Framework Convention on Climate Change (UNFCCC) opened for signature.

Negotiations on a Framework Convention had commenced in December 1990 when the UN General Assembly established the Intergovernmental Negotiation Committee for a Framework Convention on Climate Change. In five sessions more than 150 states discussed binding commitments, targets, and timetables for emission reductions, financial mechanisms, and technology transfer. During these negotiations the concept of 'common but differentiated responsibilities and respective capabilities' (CBDR-RC) was established, one of the most defining principles of UNFCCC negotiations and fundamental to operationalising the notion of equity (Blaxekjær and Nielsen 2014; Winkler and Rajamani 2014). The UNFCC was adopted in 1992 in New York, USA, alongside the UN Convention on Biological Diversity and the Convention to Combat Desertification. The treaty entered into force on March 21 1994, signed by 196 countries (also known as 'parties').

The political response to climate change in the form of the UNFCCC aims to stabilise atmospheric concentrations of greenhouse gases (GHG) to avoid 'dangerous interference with the climate system' (United Nations 1992, p. 9). Annual meetings, known as the Conference of Parties (COP), which form the basis for negotiating multilateral responses to climate change, are held at the end of each year. The following sections include (Sect. 2.2) an overview of milestones in the history of climate change negotiations, (Sect. 2.3) an insight into the multi-annual battle that defined the history of climate change negotiations with particular focus on the period between COP15 and COP21, and (Sect. 2.4) a description of the PA itself. By summarising key aspects of the 'road to Paris', the final section (Sect. 2.5) also provides a glimpse into consecutive chapters and new climate governance possibilities enabled by the PA.

2.2 A Brief History of Climate Change Negotiations

The first Conference of Parties (COP1), presided over by Germany's then environment minister Angela Merkel, was held in Berlin, Germany, December 1995. During COP1 the parties established that commitments in the Convention were 'inadequate' for meeting UNFCCC objectives. The resulting Berlin Mandate formed an agreement for the establishment of 'a protocol or another legal instrument' to strengthen the commitment of Parities included in Annex I (henceforth Annex I countries) to the UNFCCC (United Nations 1995). The process laid out by the Ad Hoc Group on the Berlin Mandate (AGBM), including the identification of appropriate actions for the period beyond 2000, laid the foundation for the Kyoto Protocol (KP).

COP3 held in Kyoto, Japan, December 1997, is widely regarded as the first significant milestone in international climate change negotiations. The resulting KP, adopted on 11 December 1997, was the world's first GHG or carbon[1] emission reductions treaty with legally binding commitments. It committed its parties to internationally binding emission reduction targets and recognised that industrialised and developed countries (Annex I countries) carry the main responsibility for atmospheric carbon concentration in the last 150 years under the principle of CBDR-RC (UNFCCC 1997). Under the KP, parties with carbon emission reduction requirements (Annex I countries) must meet their targets primarily through national carbon emission reduction measures during the first commitment period (2004–2008). The KP also provides mechanisms to help parties meet their target.

Another significant milestone in international climate change negotiations was reached during the second part of COP6 in Bonn, Germany, December 2001. It oversaw parties reaching a broad political agreement on the operational rulebook for the KP. At COP7 in Marrakesh, Morocco, in November of the same year, detailed implementation rules for the KP were adopted. This paved the way for its ratification and for formalising agreement on operational rules.

The KP entered into force on 16 February 2005 according to detailed implementation rules adopted at COP7 in Marrakesh, Morocco, December 2001. It was made possible by the Russian Federation's submission of its instrument of ratification to the KP in 2004, which ensured that the minimum requirement for the KP to come into effect (ratification by 55 industrialised countries, representing 55% of global carbon emissions) was exceeded (Clémençon 2008). The KP's first commitment period started in 2008 and ended in 2012. During this period, 37 industrialised countries and the European Union committed to reduce carbon emission by 5% on average against 1990 levels (UNFCCC 1997). 2005 was also the year in which the first Emissions Trading Scheme (ETS) in the world, the European Union ETS (EU-ETS), was launched. EU-ETS is the world's largest carbon market, operating in 31 countries (all 28 EU countries plus Iceland, Liechtenstein and Norway).

COP11 in Montreal, Canada, December 2005, was held in conjunction with the first Conference of the Parties serving as the Meeting of the Parties (CMP1) to the KP. CMP reviews the implementation of the KP and takes decisions to promote its effective implementation. One month after CMP1, the Clean Development Mechanism (CDM), a key mechanism under the KP, was launched. The CDM, which was included in the KP following an initiative by Brazil to create an incentive for financing carbon emission reduction measures in developing countries

[1] Carbon Dioxide (CO_2) is the most common Greenhouse Gas (GHG) and is often used as a reference unit to express levels of GHG emissions (expressed in CO_2 equivalent, or CO_2eq). In the context of this Chapter and, more in general, of this book the term carbon, expressing CO_2, and the term GHG are synonyms, hence carbon is used as single term to express GHG.

(Clémençon 2008), allows parties with emission reduction or limitation commitments under the KP (Annex I countries) to implement emission reduction projects in developing countries (non-Annex I countries). Emission reduction projects in non-Annex I countries can earn saleable certified emission reduction credits (CERs—equivalent to one tonne of CO_2), which can be counted towards meeting KP targets of Annex I countries. As the first global environmental investment and credit scheme of its kind, its CERs also introduced a standardised emissions offset instrument (Stua 2014).

At COP13 held in Bali, Indonesia, December 2007, the Bali Road Map was adopted. It charted the courses for a new negotiating process to address climate change according to five main categories: shared vision, mitigation, adaptation, technology, and financing. It also set out the steps 'to reach an agreement on long-term cooperative action up to and beyond 2012 to be adopted at its 15th session in Copenhagen in December 2009' (UNFCCC 2007). COP13 was also successful in progressing the Adaptation Fund (AF), in including technology transfer and avoidance of deforestation in a future agreement through REDD (Reducing Emissions from Deforestation and Degradation), and in progressing CDM (Clémençon 2008). One month after COP13, Joint Implementation (JI), another key mechanism under the KP, was launched. The JI allows countries with emission reduction or emission limitation commitments under the KP (Annex I countries) to earn emission reduction units (ERUs) from emission reduction or emission removal projects in countries with similar commitments, which can be counted towards meeting the Kyoto targets. From the onset it was assumed that JI would primarily result in the construction of eligible facilities in 'transition economies' of Eastern Europe paid for by Western European and North American countries.

At COP14 held in Poznań, Poland, December 2008, the AF was launched. Established under the KP in 2001, its purpose is to finance concrete adaptation projects and programmes in developing countries that are particularly vulnerable to the adverse effects of climate change. The AF is supervised and managed by the Adaptation Fund Board. It is financed with a share of the proceeds from CDM project activities (2% of CERs issued for a CDM project activity).

At COP15, December 2009, world leaders gathered in Copenhagen amid high expectations for far reaching agreements. The resulting Copenhagen Accord (CA) is considered a disappointment, if not a failure, by many commentators (Blaxekjær and Nielsen 2014; Christoff 2010; Grubb 2013; Hourcade and Shukla 2015; Roberts 2011) despite pledges by developed countries to provide up to USD 30 billion a year in fast-start finance for the period 2010–2012 and USD 100 billion a year by 2020. The main points of criticism were the low targets and missing goals. Regarding targets, the CA recognised the scientific case for limiting global average temperature rise to 2 °C without containing commitments to reduce emissions accordingly. Regarding missing goals, all references to 1.5 °C in drafts were dropped, as was the goal of reducing global CO_2 emissions by 80% by 2050 (Vidal et al. 2009).

At COP16 held in Cancun, Mexico, December 2010, the Green Climate Fund (GCF), the Technology Mechanism and the Cancun Adaption Framework (CAF) were established. The GCF was set up to reduce carbon emission in developing countries and to help adapt vulnerable societies to climate change using the funds pledged by Annex I countries in Copenhagen. The GCF pays particular attention to the needs of those in highly vulnerable circumstances, such as Least Developed Countries (LDCs) and Small Island Developing States (SIDS) (and the Alliance of Small Island States (AOSIS), the representative organisation of SIDS established in 1992). The Technology Mechanism supports the development and transfer of climate change mitigation technologies to developing countries and the acceleration of their sustainable development. The CAF called for adaptation to be addressed with the same level of priority as mitigation (UNFCCC 2011a).

What might appear as a consolidating COP to pick up the pieces left behind by the failure of COP15 has been interpreted by Hourcade and Shukla (2015) as a paradigm shift because COP16 oversaw the emergence of the notion of 'equitable access to sustainable development' (EASD), which enlarged the concept of equity beyond 'burden sharing' and the Annex I/Non-Annex I logic. Hourcade and Shukla (2015) argue that the deadlock of climate negotiations up to and including COP15 was the result of a misguided focus on 'fairness' of emissions' allocation in climate negotiations. In this light, COP16 moved beyond the self-defeating process of negotiations since COP1 because the new regime to emerge at COP16 focused more on capabilities than requirements.

At COP17 held in Durban, South Africa, December 2011, parties committed to a new universal climate change agreement by 2015 for the period beyond 2020, effectively an 'agreement to agree' (Grubb 2013). The period beyond 2020 marks the end of the second commitment period of the KP, which had been a critical part of the international climate regime to date. The second commitment period was therefore set to bridge the period before a new agreement would take effect. Although sometimes decried as ineffectual or symbolic (Boehringer and Vogt 2004), the KP needs to be credited for successfully committing developed countries to internationally binding emission reduction targets until a new climate agreement would take effect in 2020.

At COP17, uncertainty about the form that the replacement of the KP would take led to the launch of the Ad Hoc Working Group on the Durban Platform for Enhanced Action (ADP). The mandate of the ADP was to develop 'a protocol, another legal instrument or an agreed outcome with legal force under the Convention applicable to all parties' (UNFCCC 2011b, p. 2) to be completed no later than 2015, adopted at COP21, and to come into effect and be implemented by 2020. Many commentators consider COP17 an important turning point in international climate change negotiations (Rajamani 2014; Roberts 2016; Tollefson 2011).

At COP18 held in Doha, Qatar, December 2012, the need for a binding agreement in 2015 was further stressed and parties agreed to work towards such an agreement. They also committed to find ways to scale up efforts before 2020 beyond existing pledges to curb carbon emissions. COP18 therefore reoriented the UNFCCC process towards the 2015 deadline. With the adoption of the Doha Amendment, the parties launched a second commitment period of the KP, which ensured that Annex I countries were still bound by legally binding caps on emissions, albeit countries with a declining share of global emissions (around 15% in 2013; Grubb 2013).

Ratification of the Doha Amendment to the KP was considered important to maintain momentum for global climate action for the years leading up to 2020 (Grubb 2013). Possibly more important, however, was the extension of the adaptation levy from the CDM to JI and emissions trading through international transfers of Assigned Amount Units (AAU). According to Grubb (2013), this levy is the equivalent of a 'Tobin tax' on international carbon trading and with the Doha decision it is established in international law for countries participating in the second commitment period of the KP, which therefore sets a regime standard.

At COP19 held in Warsaw, Poland, December 2013, governments took further steps towards securing a global climate change agreement in 2015. The Warsaw Outcomes included a rulebook for reducing emissions from deforestation and forest degradation and a mechanism to address loss and damage caused by long-term climate change impacts. More importantly, it established that the objective of the 2015 agreement would 'bind nations together into an effective global effort to reduce emissions rapidly enough to chart humanity's longer-term path out of the danger zone of climate change, while building adaptation capacity' and 'to stimulate faster and broader action now' (UNFCCC 2013).

Governments also agreed to communicate their respective contributions towards the universal agreement in advance of COP21 in Paris, France. This invitation to submit Intended Nationally Determined Contributions (INDCs) marked the initiation of intensified domestic preparations towards achieving climate change mitigation objectives. At the same time, Monitoring, Reporting and Verification (MRV) arrangements for domestic action were finalised, which formed another requirement for an agreement to be reached in 2015 (Rajamani 2014; UNFCCC 2014a). MRV encourages transparency of domestic carbon reporting to ensure accurate centralised accounting.

In March 2014 the UNFCCC celebrated its 20th anniversary. In September of the same year the UN Secretary-General's Climate Summit was hosted by Ban Ki-moon in New York, USA, to mobilise ambitious action on climate change in advance of COP21 in Paris in 2015. COP20 held in Lima, Peru, in December 2014 was dominated by intense discussions in anticipation of COP21. The 'Lima Call for Climate Action' was subsequently adopted, which triggered the process for submitting and reviewing INDCs ahead of the Paris conference and called for deep cuts in carbon emissions to achieve the 2 °C target (UNFCCC 2014b). In the run-up to COP21, the submission of INDCs, which were subsequently named Nationally Determined Contributions (NDCs), provided evidence of widespread political commitment to address climate change.

At COP21 in Paris, France, December 2015, a new agreement was reached. The Paris Agreement (PA) states that parties commit to hold 'the increase in the global average temperature to well below 2 °C above pre-industrial levels and to pursue efforts to limit the temperature increase to 1.5 °C above pre-industrial levels' (UNFCCC 2015, Article 2). Meeting this primary objective is a binding target for the countries adhering to the PA.

2.3 The Multi-annual Battle Culminating in the Paris Agreement

The previous section outlined the history of climate change negotiations. However, what may at times appear as a linear progression entails an intricate multi-annual battle. This section provides an insight into key moments and turning points defining the evolution of climate negotiations with particular focus on COPs designated for binding agreements, namely COP3 in Kyoto, COP15 in Copenhagen, and COP21 in Paris. It places particular emphasis on internal factors, such as shifting alliances, changing positions, and emergent diplomatic approaches such as 'huddle diplomacy' (Tollefson 2011; Rajamani 2014) but also takes external factors, such as physical climate change, interest groups, and theoretical debates on *carbon* markets, into account.

The KP, which entered into force in February 2005, was the first agreement imposing binding legal commitments to carbon emission and was celebrated as a breakthrough in international climate change mitigation (Oberthuer and Ott 1999). It has been argued that the KP represents the first step in the ultimate design of an effective, long-term, international climate change policy architecture. The emission commitments for the KPs first commitment period (2008–2012) were among the most contentious elements negotiated at the COP3. Up to that point, a quantitative interpretation of the long-term goal of atmospheric stabilisation of carbon emission to prevent dangerous anthropogenic interference with global climate had not been reached (Aldy and Stavins 2007).

Despite the pioneering role of the KP, it failed on many accounts. Some argue it was merely symbolic, especially once the USA decided not to ratify it, and that it '[codified] more or less business-as-usual emissions and [made] compliance a rather cheap deal' (Boehringer and Vogt 2004, p. 597). Not everybody takes such a stark view but most agree that the KP was effectively a top-down, centralised, descriptive agreement (Aldy and Stavins 2007; Hourcade and Shukla 2015) as opposed to a bottom-up, decentralised, facilitative agreement (Boyd et al. 2015; Rajamani 2014).

COP13 2007 in Bali was an important step in the run-up to COP15 in Copenhagen. The negotiations themselves can be regarded as essentially procedural. What was most significant was the agreement to develop a roadmap for arriving at a post-Kyoto regime by the time of COP15 in 2009 (Clémençon 2008). Shifting alliances are a feature of post-KP diplomacy. In the run-up to COP15, the Group of 77 plus China (G77 + China), composed of almost the entirety of non-Annex I parties, demanded mitigation action by developed countries according to the scientific findings in the IPCC's 4th Assessment Report (25–40% reductions by 2020, 80–95% by 2050; IPCC 2007). This was considered unacceptable by developed countries given the changing global situation, especially with the financial crash placing additional burdens on the global economy (Roberts 2011).

As mentioned above, COP15 in Copenhagen was a failure, despite the number and variety of proposals and great expectations. The failure to reach agreement on a comprehensive climate deal at COP15 represented a significant setback in climate negotiations. On the upside, however, it encouraged parties and other political groups to consider new ways to overcome the deadlock (Blaxekjær and Nielsen 2014; Rajamani 2014; Rietig 2014). Prior to COP 15, the concept of CBDR-RC and associated political groups were primarily organised along the Annex I and non-Annex I divide but increasing complexity as a result of global economic uncertainty and organisational fragmentation started challenging the status quo and contributed to the establishment of new political groupings across this North-South divide (Blaxekjær and Nielsen 2014; Rajamani 2014; Rietig 2014).

What is generally accepted is that the COPs following COP15 can be regarded as positive progress (Robertson 2011; Rajamani 2014; Rietig 2014; Boyd et al. 2015; Hourcade and Shukla 2015). Where previously the UNFCCC's institutional architecture had prevented the development of an agreement (for an overview see Rietig 2014), a combination of factors such as an emerging 'climate justice' movement (Roberts 2011), greater appreciation of the win-win low carbon economic development narrative and the associated emergence and diffusion of climate policies across developed and developing countries (Rietig 2014), increasing experimentation with 'bottom-up' systems (Rajamani 2014), the need to solve problems of participation, effectiveness (meaningful rules) and compliance (with the rules) (Keohane and Raustiala 2009), as well as the above-mentioned 'equitable access to sustainable development' (EASD) (Hourcade and Shukla 2015) helped break the negotiation deadlock associated with the CBDR-RC, Annex I/non-Annex I and the North/South divide.

Due to lacking ambition of the KP and the failure to reach an agreement at COP15, the diplomatic process needed to change. Up to and including COP15, CBDR-RC were translated into quantitative commitments for industrialised countries. ETS, which emerged on the back of CBDR-RC, served as a vehicle for transferring funds to developing countries (Aldy and Stavins 2007) and as a means to support sustainable development (Stua 2014). However, many analysts had long considered this particular interpretation of CBDR-RC a hindrance to inclusive climate agreements as it posited COP negotiations as an adversarial rather than a cooperative exercise (Hourcade and Shukla 2015).

Climate diplomacy following the failure of COP15 got off to a difficult start. COP16 initially showed signs of near collapse before a surprising turn towards a positive conclusion, at least in procedural terms. At that point, the USA frequently joined the 'Umbrella Group' (comprising Australia, Canada, Japan, Kazakhstan, New Zealand, Norway, Russia and the Ukraine) while other major bargaining blocs were the EU and G77 + China, which numbered 134 countries (Roberts 2011). This provides evidence of the (re)positioning of political groups beyond the CBDR-RC principle that characterised the diplomatic debate between COP15 in Copenhagen and COP21 in Paris (Blaxekjær and Nielsen 2014; Hourcade and Shukla 2015). In this light, COP16 helped restore some of the UN's credibility as a negotiating body, although the actual outcome was 'a mixed bag' (Robertson 2011). The emergence of new groups was a consequence of the increasing realisation that even if developed countries reduced their emissions to zero, average global temperature increase would not be limited to 2 °C.

A traditionally important aspect in the context of international negotiations and corresponding treaties is the use of language. The UNFCCC language is a prime example of careful crafting and what might appear to be unnecessarily vague wording is often the result of complex negotiations, which ensures the maximum outcome in terms of options for climate change mitigation and adaptation with the minimum of controversy. Efforts to make the language of the UNFCCC more concrete failed as a result of power politics. Despite the careful wording, however, ratification of Agreements by national governments often failed. One key example is the USA signing but not ratifying the KP (Boehringer and Vogt 2004; Kanie et al. 2010; Roberts 2011).

Another key step that requires further explanation to help understand the evolution of international climate negotiations is the changing role of the USA. At COP3 in Kyoto, the USA was in a hegemonic position, even though the fact that no quantified emission reduction and limitation objectives were set for non-Annex I countries for the period until 2012 triggered rejection of the protocol by the USA and their refusal to ratify it (Kanie et al. 2010; Roberts 2011). By the time of COP16 in Cancun the balance of power had shifted. The USA had lost economic ground vis-à-vis China (Roberts 2011). The role of the USA also changed vis-à-vis other countries in the years following Copenhagen. The SIDS (and AOSIS) in particular benefited from US sympathy towards their cause. This helped the SIDS (and AOSIS) push through a surprisingly large share of 'their' agenda and interests, especially the 1.5 °C target and the underlying scientific argument (de Agueda and Mol 2014).

At the same time, the G77 negotiations bloc showed signs of fragmentation and the EU experienced both a weakening of internal consensus on climate change and a collapse in its external bargaining strength (Roberts 2011). The G77 were a complex negotiation bloc to start with, holding together at times and fracturing at others. Their common identity was originally derived from exclusion from the league of wealthy nations (Roberts and Parks 2006). As time progressed, however, groups within G77, especially BASIC (Brazil, South Africa, India and China), which emerged in 2009, increasingly followed their own agenda. It also became increasingly common for parties to voice their specific concerns by tailoring their statements in the negotiations to their national interests. As a consequence, a large bargaining unit such as the G77 proved increasingly unable to reflect these increasingly multidimensional concerns (Roberts 2011).

The EU's position was weakened as a result of the economic crash, internal dissensus and loss of power vis-à-vis the USA as well as emerging economies (Roberts 2011; Skovgaard 2014). This became obvious in its exclusion from the core negotiating group of the CA, made up of BASIC and the USA (Roberts 2011). EU accession since the KP implied that a large number of 'transitional economies' in Eastern Europe with low prioritisation of climate issues challenged the EU's ambitious stance on climate change. The period 2009–2011 saw a deepening division between those who considered climate policy detrimental (led by Poland) and those who considered it beneficial to economic growth (led by the UK), which resulted in disagreement over an increase of the EU's reduction target for 2020. Other actors such as France and Germany, despite rhetoric in favour of a step-up, remained undecided in 2011 (Skovgaard 2014). As a result, the EU's leadership on climate change targets gave way to a less clear cut role in climate negotiations.

Another important new development in the aftermath of Copenhagen was the emergence of new constellations. These include the BASIC group, the Climate Vulnerable Forum (CVF), the Cartagena Dialogue for Progressive Action (CD), the Durban Alliance (DA), the Mountainous Landlocked Developing Countries (MLDC), Like-Minded Developing Countries (LMDC) and the Association of Independent Latin American and Caribbean states (AILAC). CD and DA are political groups of both Annex I and non-Annex I countries. A defining feature of some of these new groups, particularly CD, DA, AILAC and CVF, was their approach to sharing responsibility for climate change beyond the North-South divide, thereby challenging the traditional interpretation of CBDR-RC (Blaxekjær and Nielsen 2014).

The emergence of the BASIC group as a serious negotiation partner coincided with the decline of the EU (Hurrell and Sengupta 2012; Skovgaard 2014). In fact, Copenhagen has been used as evidence for a more general power shift away from Europe and the USA towards emerging powers although their increasing economic power and increasing share of carbon emissions did not automatically encourage these countries to take on greater responsibility (Hurrell and Sengupta 2012). Far from a homogenous group, BASIC needs to be understood as a coalition in search of a joint identity which briefly drew together to coordinate their position in Copenhagen and the immediate aftermath but struggled to maintain the momentum thereafter (Hochstetler and Milkoreit 2014) The emergence of the BASIC group in 2009 nevertheless contributed to the emergence of a new climate regime, which saw a variety of proposals put forward with a wide variety of models (Hurrell and Sengupta 2012; Hochstetler and Milkoreit 2014).

The new climate regime, which according to some scholars (i.e. Kuik et al. 2008) became apparent as early as the KP's entry into force, appeared to favour mitigation (as opposed to adaptation) based on a regime of strong international law (as opposed to unilateral or small-party systems) and market mechanisms (as opposed to regulation) with incentive structures tending to favour 'sticks' (emphasising emission reductions) as opposed to 'carrots' (encouraging technological solutions) (for an overview see Kuik et al. 2008). Where the KP, as mentioned before, was effectively top-down, descriptive, centralised and binding (Aldy and Stavins 2007; Boyd et al. 2015; Hourcade and Shukla 2015), the new climate negotiation regime placed increasing emphasis on a bottom-up, facilitative, decentralised and non-binding system (Boyd et al. 2015; Rajamani 2014).

Key to understanding the emergent climate regime, which found its preliminary consolidation in the PA, is the evolution of CRDB-RC. When the UNFCCC was negotiated it made sense to distinguish between two groups of countries. Over time, however, the original differentiation between 'developed' and 'developing' countries painted a decreasingly accurate picture of reality as many countries categorised as 'developing' under the UNFCCC in 1992 saw per-capita production and CO_2 emission levels rise above the OECD mean (Michaelowa and Michaelowa 2015).

This 'ideological lock-in' (Ipsen et al. 2001; Stua 2014) with CBDR-RC being interpreted as a binary differentiation between Annex I and non-Annex I countries was evident up to COP13 in Bali (Rajamani 2013). Since then, however, interpretation of the CBDR-RC principle has gradually changed. The agreement to develop 'a protocol, another legal instrument or an agreed outcome with legal force under the Convention applicable to all parties' (UNFCCC 2011b, p. 2), at COP17 in Durban most clearly reflects this trend (Rajamani 2013).

The Durban Platform is noteworthy because the term CBDR-RC was omitted in favour of the reference 'under the Convention', which implicitly included its principles. This is seen as evidence for the recognition that the CBDR-RC principle required reinterpretation in a more nuanced fashion to help operationalize equity and help guide the fashioning of a commitment in a regime 'applicable to all' (Winkler and Rajamani 2014, p. 103). As might be expected, this extension of mitigation responsibilities to non-Annex I countries was initially strongly opposed by groups such as the G77 (Michaelowa and Michaelowa 2015). Somewhat surprisingly, given their relatively small combined population, was the important role played by AOSIS in convincing other non-Annex I countries to adopt the Durban Platform (Rajamani 2013). AOSIS also appears to have gained strength within G77 in the run-up to Paris and by the time of COP18 in Doha, even BASIC agreed on a more nuanced division of countries than the simple binary divide stating that 'all countries should participate in an enhanced global effort' along with the focus on CBDR-RC (Michaelowa and Michaelowa 2015.

The increasingly evident heterogeneity within the BASIC group became more evident at COP19 in Warsaw, where the recently formed LMDC coalition, which comprised among others, China and India (but notably neither Brazil nor South Africa), resisted the provision of detailed information by parties and argued that informational requirements as well as rules, if any, should be differentially applied along the lines of CBDR-RC (Rajamani 2014; UNFCCC 2014c). The reinterpretation of CBDR-RC was nevertheless underway, with the 'Umbrella Group' including the USA, along with the majority of other developed countries, refusing to accept any differential application in the post-2020 period. At the same time, the EU, the 'Umbrella Group' and the USA called for top-down requirements for transparency. Combined with a rigorous multilateral assessment process favoured by the AILAC and the EU, it became evident that a 'hybrid model' was emerging with a new interpretation of CBDR-RC (Boyd et al. 2015; Rajamani 2014).

Contrasting and contentious views on equity and differentiation have been essential elements in shaping this model. Some parties adamantly defended the strict differentiation according to the UNFCCC Annexes while an increasing share of parties sought to bypass the Annexes. On one hand, the LDMCs, among others, called for a prescriptive, top-down architecture for the 2015 agreement. On the other hand, the 'Umbrella Group', among others, proposed a bottom-up system based on nationally determined mitigation actions for the 2015 agreement taking national circumstances into account (Rajamani 2014; Rietig 2014).

Other parties such as the EU and the AILAC called for participation by all parties in accordance with their evolving responsibilities and capacities without the legal commitment but with a robust international assessment or review process, thereby reflecting a 'hybrid model'. Another 'hybrid model' was put forward by the Africa Group. This model included legally binding commitments for all parties although it differentiates between emission reductions for Annex I countries and mitigation actions that support a deviation from business-as-usual for non-Annex I countries, while allowing for self-differentiation with respect to specific commitments, as well as an international process for assessment, review and possible ratchet (termed 'equity and differentiation' by Rajamani 2014, p. 733).

The interplay of potential architectures posed difficulties for some negotiating interests and coalitions. The LDMCs favoured explicit differentiation in favour of non-Annex I countries but this stood in contrast with the need for the future Agreement to be 'applicable to all' (UNFCCC 2011b, p. 1). At the same time, some non-Annex I countries such as South Africa and Indonesia supported binding commitments prior to COP21 while other countries such as China and Singapore increasingly engaged in domestic mitigation action despite opposing binding commitments. Where the CBDR-RC principle had previously appeared to be used as a justification to resist demands for commitments on the part of non-Annex I countries, it became increasingly evident that even countries constantly opposing mitigation commitments within the framework of the UNFCCC were implementing significant mitigation action domestically (Michaelowa and Michaelowa 2015). Further contentious issues at Warsaw were the concepts of 'commitments' (for developed countries) vs 'actions' (for developing countries). The final decision led to the substitution of the term 'commitment' with the term 'contributions' (a prime example of careful wording). To sum up, COP19 in Warsaw played an important role in bypassing the traditional Annexes grouping to take account of national autonomy and changing economic circumstances (Rajamani 2014).

Another significant change in the run-up to Paris was the change in negotiation style. By 2013 reinforcing dynamics at the core of the UNFCCC associated with the two working streams Ad Hoc Working Group on Further Commitments for Annex I parties under the Kyoto Protocol (AWG-KP) and Ad Hoc Working Group on Long-term Cooperative Action under the Convention (AWG-LCA) had been replaced by the ADP towards a 2015 agreement (see above Sect. 2.2). The negotiation process was also changed from the traditional 'core event' structure with parallel work streams in 'side events' towards an open workshop and roundtable structure. This helped breach the divide between countries with and without obligations under the UNFCCC to set up climate legislation by encouraging developing countries to present their actions and ambitions. As a result, the question shifted from 'whether' developing countries should or should not reduce their emissions towards 'how' they can engage in low carbon development (Rietig 2014).

In practice, COP19 in Warsaw put the negotiation process that culminated in the PA on track for a hybrid architecture, combining bottom-up, decentralised, non-binding elements such as INDCs with top-down, centralised, binding elements such as rules on transparency and accounting (Boyd et al. 2015; Rajamani 2014). This was enabled by the transformation of the dominant interpretation of CBDR-RC from a binary differentiation between Annex I and non-Annex I countries towards a more nuanced interpretation and expectation that a new climate regime must contain greater symmetry in the commitments undertaken by all parties (Rajamani 2013; Winkler and Rajamani 2014). This new interpretation of CBDR-RC has been succinctly summed up by Winkler and Rajamani (2014, p. 102): 'All must do more, while some must do more still than others'.

The call on nations to communicate INDCs as introduced at COP20, which were required to be quantified, fair and ambitious in light of national circumstances and appropriate towards achieving the 2 °C target, can be seen as final proof that the process involving the old interpretation of CBDR-RC which had led to the KP had finally been supplanted by a new climate negotiation regime (Rajamani 2014; Hourcade and Shukla 2015; Rajamani 2016). The invitation to submit INDCs well in advance of COP21 in Paris facilitated the negotiation process by leaving common objectives such as a long-term vision, procedural issues relating to review cycle, MRV guidelines, financial and technology support, and provisions for establishing a carbon market on the negotiating table (Clémençon 2016; Stavins 2016).

Apart from the internal factors of the diplomatic negotiation procedure, several important external factors became increasingly important in the run-up to COP21. They can be categorised into the above-mentioned areas of physical climate change, theoretical debates on carbon markets, and the increasingly relevant role of non-party, public and private stakeholders.

Supported by the multi-annual scientific efforts of the IPCC (2007, 2014), physical evidence of climate change became more apparent in the weeks before COP21 commenced in Paris on 30 November 2015. Three key elements were registered by that time: (a) 2015 was on track to be the hottest year on record—later confirmed by NASA (2016) and Met Office (2016); (b) the planet's average surface temperature had risen 1 °C since the late-nineteenth century (NASA 2016; Met Office 2016); (c) 2016 would be the first year in which the concentration of atmospheric CO_2 would exceed 400 ppm—confirmed by WMO (2016). It is unclear exactly how much physical evidence contributed to the diplomatic debate but these figures are likely to have highlighted the urgency of the task ahead and the need to concentrate efforts on reaching a binding treaty.

Theoretical debates in the run-up to Paris have often turned conventional thinking on its head. One example is the concept of 'unburnable carbon'. It converts the requirement to limit atmospheric carbon concentrations into total burnable carbon reserves, thereby providing carbon budgets for different geographical areas and countries (McGlade and Ekins 2015; Carbon Tracker 2016). Another example is monetary innovation, which, despite its significance, has been slow to emerge as a topic of interest in the context of climate change. Lietaer et al. (2012) suggest that the missing link in models to solving overarching dilemmas such as climate change is monetary innovation and several models have emerged in recent years.[2]

The relevance of non-party, public and private stakeholders in international climate change negotiations has grown in parallel with the complexity of global

[2] Monetary innovation represents a core subject of this book and its significance is widely discussed in Chap. 9.

climate governance (Roberts 2011; Bulkeley et al. 2012; Busby and Hadden 2014; Nasiritousi et al. 2016). Initially in a predominantly adversarial role before 'the brutal arithmetic of climate change' (Nicholas Stern at COP18; Harvey 2012) became increasingly evident, many non-party, public and private stakeholders (as well as governments) tended to avoid responsibility. Over a long period of time, the actions of many of these stakeholders displayed a fear of losing privileges. The lack of support for a clear and robust climate agreement consequently represented a persistent problem in climate change negotiations. In recent years, however, an increasing number of stakeholders have started to recognise the socio-economic benefits of reducing carbon emission and engaging more broadly in sustainable development (Stua 2014).

Non-party stakeholders can be differentiated according to the following governance profiles: business and non-governmental organisations (BINGOs); environmental non-governmental organisations (ENGOs); indigenous peoples' organisation (IPOs); local government and municipal authorities (LGMAs); research and independent non-governmental organisations (RINGOs); trade unions non-governmental organisations (TUNGOs); farmers and agricultural NGOs; women and gender; and youth NGOs (YOUNGOs) (Nasiritousi et al. 2016). This list is not exhaustive and these groups themselves are far from homogenous, representing broad and sometimes conflicting interests that often fail to acknowledge the voices of non-accredited organisations (Nairitousi et al. 2016).

The governance profile of BINGOs is perceived as having a strong focus on influencing the policy-makers and the overall agenda, that of LGMAs as active in mitigation and adaptation, that of RINGOs as focusing on the proposition of solutions, provision of expertise and evaluation of consequences, while that of ENGOs is perceived as contributing strongly to awareness raising and representation of public opinion. BINGOs are closely associated with leverage and material powers compared to the cognitive powers of RINGOs, symbolic and social powers of ENGOs and implementation powers of LGMAs (Nasiritousi et al. 2016).

What is notable is the relative decline of ENGOs vis-à-vis other non-state actors. ENGOs constituted 65% of registrants in 1995–1997, 61% in 1998–2000, 58% in 2001–2006 and 46% in 2007–2009 (Busby and Hadden 2014). This relative decline, however, should not be interpreted as declining influence. ENGOs along with environmental and social movements as well as academic and policy-making publications supported and continue to support the demands of activists and developing country governments. This movement established itself under the broad heading of 'climate justice' in the wake of Kyoto, where nearly all developing countries were excluded from negotiations, and gained momentum in the run-up to Copenhagen with its focus on vulnerability, 'climate debt' (historical responsibility of carbon emissions), and the science of the IPCC reports (Roberts 2011). Despite the failure to reach an agreement at COP15, the demands of the climate justice movement have become increasingly widely accepted and can now be recognised as a driving force for the 1.5C ambition of the PA, which was championed most vocally and successfully by SIDS and their umbrella organisation AOSIS (de Agueda and Mol 2014).

At the same time, BINGOs also became more active players in global climate governance. In the run-up to COP21, for example, the so-called New York Declaration on Forests is noteworthy because it focuses on business supply chains with more than 30 governmental signatories, more than 40 business signatories and over 60 ENGOs and indigenous organisations (Busby and Hadden 2014). LGMAs and cities in particular are also increasingly recognised as key players in the fight against climate change. Important local government and city networks in this context are C40, a network of more than 80 megacities founded in 2005; ICLEI-Local Governments for Sustainability, a global network of more than 1500 cities, towns and regions founded in 1990; UCLG (United Cities and Local Governments), which represents over 240,000 towns, cities, regions and metropolises as well as 175 local and regional government associations created in 2004 as an inheritor of the International Municipal Movement; the Covenant of Mayors for Climate & Energy, an EU initiative targeting local authorities with more than 6000 signatories founded in 2008; and the Compact of Mayors, launched by the then UN Secretary-General Ban Ki-moon under the leadership of C40, ICLEI and UCLG, which specifically focused on demonstrating commitment to an ambitious global climate solution in the run-up to COP21.

2.4 The Paris Agreement

This section provides an overview of the content of the PA, some insights into preliminary interpretations and a brief synthesis of the negotiation process. The PA is 31 pages long with 16 introductory clauses and 29 operative clauses. It was formally adopted by the parties to the UNFCCC as a decision on 12 December 2015. Expectations in the PA are high. According to former UN Secretary-General Ban Ki-moon, the PA 'is a monumental triumph for people and planet. [...] It sets the stage for progress in ending poverty, strengthening peace and ensuring a life of dignity and opportunity for all' (Ki-moon 2015).

By charting a new political course for global climate change action it 'provides the first real hope that we might be able to address climate change in a more timeous, responsible, united, equitable and sustainable way' (Roberts 2016, p. 1). What sets COP21 apart from previous COPs is that the significance of CBDR-RC has radically changed (Clémençon 2016; Stavins 2016). Although CBDR-RC maintains its importance as a guiding principle of the PA, the strict binary interpretation of CBDR-RC along the lines of Annex I and non-Annex I parties has been replaced by 'a nuanced form of differentiation between developed and developing countries' (Rajamani 2016, p. 1), which recognises that everyone needs to act according to their respective capabilities and resources to tackle climate change.

Reinterpreting the meaning of CBDR-RC beyond the differentiation between developed and developing countries and omitting references to the dominant Annex-I/non-Annex-I narrative marks a significant break from the past and the commencement of a new climate regime. It has been condemned by some commentators as an imposition of potential additional burdens on developing countries whose 'future economic options, particularly in terms of energy, will be sharply curtailed for the medium- and long-term future' (Jayaraman and Kanitkar 2016, p. 10). Similarly, the lack of equitable burden-sharing requirements among parties to control and reduce carbon emission fails to both prioritise 'equity and environmental justice considerations as a guideline for multilateral cooperation' and 'provide a blueprint for achieving [...] stabilisation objectives' (Clémençon 2016, p. 3). Others have hailed this reinterpretation as an important step towards a universal agreement on climate change because the PA involves all countries in carbon emission mitigation (Marcu 2016; Stavins 2016; Stua 2016).

The USA is often held responsible for the exclusion of binding emission targets and timetables for each party. However, for domestic reasons, the USA was never in a position to accept a climate agreement formally based on binding recognition of past emissions. 'It therefore advocated for a universal global agreement that would include binding provisions for developing countries as well' (Clémençon 2016, p. 5). In this context the PA may be interpreted as a nod towards the complicated domestic political situation of the USA where a Congress majority does not believe in human responsibility for climate change (Clémençon 2016; Jayaraman and Kanitkar 2016).

Some of the key Articles of the PA are outlined and briefly described below.

Article 2: This article reaffirms the primary objective of holding 'the increase in the global average temperature to well below 2 °C above pre-industrial levels and to pursuing efforts to limit the temperature increase to 1.5 °C above pre-industrial levels' (UNFCCC 2015, p. 2/22). The 1.5 °C objective in particular is a much stronger outcome than anticipated although the difficulty of reaching this target is evident from calculations provided by the IPCC (2007, 2014) and the United Nations Environment Programme (UNEP 2014). The 1.5 °C target can be seen as a success of the SIDS (and AOSIS) in putting forward the scientific argument, in getting powerful developed countries on board and in framing their domestic mitigation activities according to the 1.5 °C target (de Agueda Corneloup and Mol 2014).

According to the IPCC's 5th Assessment Report (IPCC 2013/2014, p. 63), the world can emit an additional 1000 $GtCO_2$ until 2100 if it wants to limit global average temperature rise to 2 °C above pre-industrial levels with a probability of > 66%. In 2010, anthropogenic carbon emissions reached 49 ± 4.5 $GtCO_2e$ (IPCC 2014, p. 5). At current growth rates, the 1000 $GtCO_2$ allowance would be used up shortly after 2030. According to UNEP (2014), global emissions should not exceed 42 $GtCO_2e$ in 2030 if 2 °C is the target and 39 $GtCO_2e$ if it is 1.5 °C, with a probability of > 66% that temperatures will remain below target values. To have a 50% chance of keeping global warming below 1.5 °C, carbon emissions would have to fall from a projected 56 $GtCO_2e$ in 2020 to 8 by 2050 (UNEP 2014).

Article 2 also reaffirms the CBDR-RC principle by stating that the Agreement will be implemented to reflect equity as well as the CBDR-RC principle (UNFCCC 2015, p. 22). What sets the PA apart is the inclusion of the clause 'in light of different national circumstances', which was first explicitly referred to in the Lima Call for Climate Action in 2014 (Rajamani 2016). Arguably, the PA does not provide guidance on the operationalisation of equity (Jayaraman and Kanitkar 2016). However, another interpretation of this paragraph, taking Article 4 into account (see below), which states that 'each party's successive nationally determined contribution will represent a progression beyond the Party's then current national determined contribution and reflect its highest possible ambition, reflecting the CBDR-RC principle', suggests that it forms the basis for a strong (but not binding) obligation for each party to take more ambitious actions over time (Rajamani 2016).

Article 4: This article states the long-term objective 'to achieve a balance between anthropogenic emissions by sources and removals by sinks of greenhouse gases in the second half of this century' (UNFCCC 2015, p. 22). Given the global warming ceiling specified in Article 2, most scientists interpret this to mean that global net carbon emissions need to reach zero by 2060–2080 (Clémençon 2016). This also implies that beyond the achievement of Net Zero Carbon (NZC) emissions, carbon needs to be actively removed from the atmosphere.

The article also states that all countries should 'strive to formulate and communicate long-term low greenhouse gas emission development strategies' (UNFCCC 2015, p. 23). In combination with binding updates every 5 years, this implies that collective ambitions will be ratcheted up over time as each successive step needs to be at least as strong as the current one (Clémençon 2016; Stavins 2016; UNFCCC 2015, p. 22). This implies that parties have to submit new or renewed pledges for 2020 9–12 months before the 2020 COP meeting. The domestic process for establishing new pledges would need to start no later than 2018 (UNFCCC 2015, p. 4).

Article 6: This article is the most significant innovation in international climate negotiations and provides the foundation for the entire proposal introduced in this book. Interpreted by many as the Article for carbon markets, it encompasses a much wider perspective by providing the framework for implementation of new, experimental forms of climate governance. Chapter 4 of this book is entirely dedicated to Article 6, including a full analysis and interpretation of its significance.

Article 7: This article points towards the desire to establish a global goal on 'enhancing adaptive capacity, strengthening resilience and reducing vulnerability to climate change' (UNFCCC 2015, p. 25). This article stresses the differentiated focus of developing countries on adaptation as opposed to the developed countries' focus on mitigation while linking the amount of adaptation and its cost to the level of mitigation action. The recognition of adaptation and mitigation as equally important is unique in the history of climate agreements (Roberts 2016). This article also makes specific reference to gender-responsive models and the valuation, 'as appropriate', of traditional knowledge, knowledge of indigenous peoples and 'local knowledge systems' with reference to the integration of such knowledge in relevant socioeconomic and environmental policies (UNFCCC 2015, p. 25).

Article 8: This article arguably puts loss and damage on equal footing with mitigation and adaptation, which may be interpreted as recognition of the demands of small islands and other highly vulnerable countries to climate change. However, the above-mentioned omission of any liability or compensation requirements removes the possibility of climate reparation claims arising from developed countries' responsibilities, given that historically they have contributed significantly more to atmospheric carbon accumulation than developing countries (Blanco 2014; Roberts 2016; WRI 2014).

Article 9: This article states that 'developed country parties shall provide financial resources to assist developing country parties with respect to both mitigation and adaptation in continuation of their existing obligations under the Convention' (UNFCCC 2015, p. 26). It thereby reinforces earlier agreements, specifically the CA, which stated that the developed countries collectively commit 'to provide new and additional resources [...] approaching USD 30 billion for the period 2010–2012' and 'USD 100 billion dollars a year by 2020 to address the needs of developing countries' (UNFCCC 2009, p. 7).

Article 13: This article recognises the importance of transparency to promote effective implementation in order 'to provide a clear understanding of climate change action in light of the objective of the Convention as set out in its Article 2' (UNFCCC 2015, p. 28). Transparency through monitoring, reporting and verification (MRV), has been interpreted as binding for all parties with respect to all elements contained in the PA (Bodansky 2016).

Article 14: This article introduces the concept of a 'global stocktake'. Following on from Article 13, parties commit to providing comprehensive, facilitative and transparent periodic inventories to assess collective progress towards achieving the purpose of the PA and its long-term goals. The COP also commits to taking its first global stocktake in 2023 and every 5 years thereafter (UNFCCC 2015, p. 29). This is likely to be in recognition of the INDCs' shortfall in terms of limiting global average temperature to well below 2 °C above pre-industrial levels, given that, at best, they point towards a global average temperature increase of 2.7 °C if not 3 °C by 2100 (Clémençon 2016; Roberts 2016). This article implies that collective efforts will be assessed in 2018 to help update and enhance individual country plans to provide the basis for the first global stocktake in 2023.

Article 15: This article states that 'a mechanism to facilitate implementation of and promote compliance with the provisions of this Agreement is hereby established' (UNFCCC 2015, p. 29). Following paragraphs emphasise that the mechanism shall consist of an expert-based, facilitative committee and function transparently, non-adversarially and non-punitively with particular attention to respective national capabilities and circumstances of parties. It will serve as a meeting of the parties to the PA at its first session and will then report annually to the COP serving as the meeting of the parties to the PA.

2.5 Conclusions: Towards an Experimental Climate Governance

The long road towards finding a global response to climate change finally resulted in a momentous milestone in international politics, the PA. Yet, several challenges await the effective implementation of the Agreement's ambitious objectives. It is evident from the PA that the diplomats tried to reconcile the lack of ambition within the INDCs with the need to limit carbon emissions further to remain within a reasonable chance of limiting global average temperature rise to 2 °C. As the UNFCCC notes with concern (UNFCCC 2015, 2016b), the INDCs submitted to the UNFCCC are not sufficient to hold the increase in global average temperature to below 2 °C above pre-industrial levels, instead leading to a projected level of 55Gt of carbon emissions in 2030 although no more than 40Gt of carbon emissions may be released in 2030 to limit global average temperature rise to 2 °C (UNFCCC 2015, 2016b).

On a more positive note, the negotiations surrounding the (market and non-market) approaches to mitigation—which were among the last to be finalised on the last night of COP21 before the text was finally approved by the COP21 president, French Foreign Minister Laurent Fabius—resulted in a text, which offers wide ranging possibilities (Marcu 2016; Stua 2016). The origins of this text lie in a number of submissions prior to COP 21, including those of Brazil (November 2014); Japan (September 2015); AOSIS (December 2015); EU-Brazil (December 2015); LMDC (December 2015); and Panama (December 2015) (Marcu 2016). The core focus of this book, Article 6 of the PA, owes much to these submissions. The Brazilian submission of 10 November 2014 (Brazil 2014; Marcu 2016; Stua 2016) is of particular interest here because it introduces the concept of 'Economic Instruments' that features prominently in Article 6 of the PA:

The Economic Mechanism shall be comprised of general guidelines related to an ETS and an enhanced Clean Development Mechanism (CDM+). (…) The new market mechanism (…) should be established under the agreement, incorporating the modalities, procedures and methodologies of the Clean Development Mechanism, to allow trading of CER among all parties (Brazil 2014, p. 11–12).

Building upon Article 6, its history and the possibilities for its interpretation as a tool for the implementation of new, experimental climate governance regimes, this book opens a unique window of opportunity for a real, economically viable and socially acceptable low-carbon transformation.

References

Aldy, J., & Stavins, R. N. (2007). Architectures for an international global climate change agreement: Lessons for the international policy community. In J. Aldy & R. N. Stavins (Eds.), *Architectures for agreement: Addressing global climate change in the post-Kyoto world* (pp. 350–367). Cambridge: Cambridge University Press.

Blanco, G. (2014). *Climate change 2014: Mitigation of climate change – Trends in stocks and flows of GHGs and their drivers*. Working group III contribution to the IPCC 5th assessment report.

Blaxekjær, L., & Nielsen, T. (2014). Mapping the narrative position of new political groups under the UNFCCC. *Climate Policy, 15*(6), 751–766.

Bodansky, D. (2016). The legal character of the Paris Agreement. *Review of European Community and International Environmental Law, 25*(2), 142–150.

Boehringer, C., & Vogt, C. (2004). The dismantling of a breakthrough: The Kyoto Protocol as symbolic policy. *European Journal of Political Economy, 20*, 597–617.

Boyd, R., Green, F., & Stern, N. (2015). *The road to Paris and beyond.* London: Centre for Climate Change Economics and Policy & Grantham Research Institute on Climate Change and the Environment.

Brazil (2014, November). *Views of Brazil on the elements of the new agreement under the convention applicable to all parties*. Submission by Brazil to UNFCCC.

Bulkeley, H., Andonova, L., Backstrand, K., Betsill, M., Compagnon, D., Duffy, R., Kolk, A., Hoffmann, M., Levy, D., Newell, P., Milledge, T., Paterson, M., Pattberg, P., & VaDeveer, S. (2012). Governing climate change transnationally: Assessing the evidence from a database of sixty initiatives. *Environment and Planning Government and Policy, 30*(4), 591–612.

Busby, J., & Hadden, J. (2014). *Nonstate actors in the climate arena.* The Stanley foundation working paper.

Carbon Tracker. (2016). What does 'unburnable carbon really mean? *Carbon Tracker News.*

Christoff, P. (2010). Cold climate in Copenhagen: China and the United States at COP15. *Environmental Politics, 19*(4), 637–656.

Clémençon, R. (2008). The Bali road map: A first step on the difficult journey to a post-Kyoto Protocol agreement. *Journal of Environment & Development, 17*(1), 70–94.

Clémençon, R. (2016). The two sides of the Paris Agreement: Dismal failure or historic breakthrough? *Journal of Environment and Development, 25*(1), 3–24.

De Agueda Corneloup, I., & Mol, A. (2014). Small island developing states and the international climate change negotiations: The power of moral "leadership". *International Environmental Agreements, 14*, 281–297.

Grubb, M. (2013). Doha's dawn? *Climate Policy, 13*(3), 281–284.

Harvey, F. (2012, December 4). Lord stern: Developing countries must make deeper emissions cuts. *The Guardian.*

Hochstetler, K., & Milkoreit, M. (2014). Emerging powers in the climate negotiations: Shifting identity conceptions. *Political Research Quarterly, 67*(1), 224–235.

Hourcade, J-C., & Shukla, P. R. (2015). Cancun's paradigm shift and COP 21: To go beyond rhetoric. *International Environmental Agreements, 15*(4), 343–351.

Hurrell, A., & Sengupta, S. (2012). Emerging powers, north-south relations and global climate politics. *International Affairs, 88*(3), 463–484.

IPCC. (2007). *Climate change 2007: Synthesis report.* IPCC Reports, Geneva, Switzerland.

IPCC. (2014). *Climate change 2014: Synthesis report.* IPCC Reports, Geneva, Switzerland.

IPCC. (2016). *History.* https://www.ipcc.ch/organization/organization_history.shtml

Ipsen, D., Rosh, R., & Scheffran, J. (2001). Cooperation in global climate policy: Potentialities and limitations. *Energy Policy, 29*, 315–326.

Jayaraman, T., & Kanitkar, T. (2016). The Paris Agreement – Deepening the climate crisis. *KPW Economic & Political Weekly, 51*(3), 10–13.

Kanie, N., Nishimoto, H., Hijioka, Y., & Kameyama, Y. (2010). Allocation and architecture in climate governance beyond Kyoto: Lessons from interdisciplinary research on target setting. *International Environmental Agreements, 10*, 299–315.

Keohane, R., & Raustiala, K. (2009). Toward a post-Kyoto climate change architecture: A political analysis. In *Post Kyoto international climate policy: Implementing architectures for agreement* (pp. 372–402). New York: Cambridge University Press.

Ki-moon, B. (2015). UN chief hails new climate change agreement as 'monumental triumph'. *UN News Centre.*

Kuik, O., Aerts, J., Berkhout, F., Biermann, F., Bruggink, J., Gupta, J., & Tol, R. (2008). Post-2012 climate policy dilemmas: A review of proposals. *Climate Policy, 8*(3), 317–336.

Lietaer, B., Arnsperger, C., Goerner, S., & Brunnhuber, S. (2012). *Money and sustainability – The missing link.* A report from the club of Rome – EU chapter to finance watch and the World Business Academy.

Marcu, A. (2016). *Carbon market provisions in the Paris Agreement (article 6)* (CEPS special report, 128).

McGlade, C., & Ekins, P. (2015). The geographical distribution of fossil fuels unused when limiting global warming to 2 °C. *Nature, 517*, 187–190.

Metoffice. (2016). *2015: The warmest year of record, say scientists.* United Kingdom Meteorological Office News, 2015.

Michaelowa, A., & Michaelowa, K. (2015). Do rapidly developing countries take up new responsibilities for climate change mitigation? *Climate Change, 133*, 499–510.

NASA. (2016). *NASA, NOAA analyses reveal record-shattering global warm temperatures in 2015.* NASA Press Release.

Nasiritousi, N., Hjerpe, M., & Linner, B. O. (2016). The roles of non-state actors in climate governance: Understanding agency through governance profiles. *International Environmental Agreements, 16*, 109–126.

Oberthuer, S., & Ott, H. (1999). *The Kyoto Protocol: International climate policy for the 21st century*. Berlin: Springer.

Rajamani, L. (2013). Differentiation in the emerging climate regime. *Theoretical Inquiries in Law, 14*, 151–171.

Rajamani, L. (2014). The Warsaw climate negotiations: Emerging understandings and battle lines on the road to the 2015 climate agreement. *International and Comparative Law Quarterly, 63*(3), 721–740.

Rajamani, L. (2016). Ambition and differentiation in the 2015 Paris Agreement: Interpretative possibilities and underlying politics. *International and Comparative Law Quarterly, 65*(2), 493–514.

Rietig, K. (2014). Reinforcement of multilevel governance dynamics: Creating momentum for increasing ambitions in international climate negotiations. *International Environmental Agreements, 14*, 371–389.

Roberts, J. T. (2011). Multipolarity and the new world (dis)order: US hegemonic decline and the fragmentation of the global climate regime. *Global Environmental Change, 21*, 776–784.

Roberts, D. (2016). A global roadmap for climate change action: From COP17 in Durban to COP21 in Paris. *South African Journal of Science, 112*(5/6). doi: http://dx.doi.org/10.17159/sajs.2016/a0158.

Roberts, J. T., & Parks, B. (2006). A model of north-south (non)cooperation. In B. Parks & J. T. Roberts (Eds.), *A climate of injustice: Global inequality north-south politics, and climate policy* (pp. 25–66). Cambridge: MIT Press.

Robertson, H. (2011, January 26). Cancun: A first step to replacing Kyoto. *Petroleum Economist, 78*(1), 22

Skovgaard, J. (2014). EU climate policy after the crisis. *Environmental Politics, 23*(1), 1–17.

Stavins, R. N. (2016). Trade Talks Can Follow Path of Paris Climate Change Agreement. *The Environmental Forum, 33*, 15.

Stua, M. (2014, September). *A new paradigm: Establishing a global carbon market as an element for the foundation of a 'Low Carbon Bretton Woods' system*. SPRU Working Paper.

Stua, M. (2016, March). *Establishing a roadmap for the low-carbon transformation of the dominating socio-economic systems – A review of the mechanism proposed in Article 6 of the Paris Agreement*. Social Science Research Network (SSRN) Working Paper.

Taylor, A. (2011). *The dance of air and sea*. Oxford: Oxford University Press.

Tollefson, J. (2011). Durban maps path to climate treaty. *Nature, 480*(7377), 299–300.

UNEP. (2014). *The emissions gap report 2014*. United Nations Environment Programme (UNEP) report, Nairobi, Kenya.

UNFCCC. (1997). *Kyoto Protocol to the United Nations framework convention on climate change*. Bonn: Climate Change Secretariat.

UNFCCC. (2007). *Decision –/CP.13, Bali action plan*. Bonn: Climate Change Secretariat.

UNFCCC. (2009). *Report of the Conference of the Parties on its fifteenth session, held in Copenhagen from 7 to 19 December 2009*. Bonn: Climate Change Secretariat.

UNFCCC. (2011a). *Report of the conference of the parties on its sixteenth session*. Bonn: Climate Change Secretariat.

UNFCCC. (2011b). *Report of the conference of the parties on its seventeenth session*. Bonn: Climate Change Secretariat.
UNFCCC. (2013). *Warsaw outcomes*. Bonn: Climate Change Secretariat.
UNFCCC. (2014a). *Intended Nationally Determined Contributions (INDCs)*. Bonn: Climate Change Secretariat.
UNFCCC. (2014b). *Lima call for climate action*. Bonn: Climate Change Secretariat.
UNFCCC. (2014c). *Submission by China: China's submission on the work of the ad hoc working group on Durban platform of enhanced action, 6 march 2014*. Bonn: Climate Change Secretariat.
UNFCCC. (2015). *Adoption of the Paris Agreement*. Bonn: Climate Change Secretariat.
UNFCCC. (2016). *Aggregate effect of the intended nationally determined contributions: An update*. Bonn: Climate Change Secretariat.
United Nations. (1992). *United Nations framework convention on climate change*. Rio de Janeiro, Brazil.
United Nations. (1995). *The Berlin Mandate*. Berlin, Germany.
Vidal, J., Stratton, A., & Goldenberg, S. (2009, December 18). Low targets, goals dropped: Copenhagen ends in failure. *The Guardian*.
Winkler, H., & Rajamani, L. (2014). CBDR&RC in a regime applicable to all. *Climate Policy, 14*(1), 102–121.
WMO. (2016). CO2 breaches milestone, drives warming. *World Meteorogical Organization (WMO) News*.
WRI. (2014). *The history of carbon dioxide emissions*. World Resources Institute (WRI) report.
Zillman, J. (2009). A history of climate activities. *WMO Bulletin, 58*(3), 141.

Chapter 3
Climate Clubs and Their Relevance Within the Paris Agreement

Abstract As diplomatic crisis threatened to overwhelm the negotiations for an international regime on climate change, certain new theoretical climate policy models began to gain a foothold in international debate. Hypotheses for smaller scale negotiations based upon a reduced number of actors, as well as the idea of 'minilateral' clubs of cooperating countries taking the lead in acting against climate change, quickly spread among scholars and policy makers, leading several commentators to view them as possible 'last stance' solutions in case the international climate negotiations failed. The Paris Agreement breakthrough, hailed by most as the ultimate victory of 'multilateralism', eventually overtook any minilateral ambition, at least concerning the climate bargaining process. Nevertheless, the high degree of uncertainty surrounding the implementation of the Paris Agreement's ambitious objectives may represent a window of opportunity for a specific category of 'minilateral' models, known as 'transformational clubs', to find their own significant space within the new international climate regime. While describing the general characteristics of climate clubs and the specific features of 'transformational clubs', this Chapter discusses to what extent these latter may effectively play a role in the future developments of the international debate on climate change.

Keywords Climate clubs • Transformational clubs • International regime • Plurilateral frameworks • Experimentalist governance (XG) • Paris Agreement (PA)

3.1 Introduction

During the most delicate and uncertain phase of international climate negotiations, especially the years immediately following the Copenhagen failure in 2009, a plethora of climate frameworks alternative to 'multilateralism'[1] emerged.[2] These 'minilateral' frameworks were put forward to address and help overcome the stalemate mainly caused by friction between developing and developed countries. The promoters of these frameworks, better known as climate clubs, theorised that shifting

[1] In the context of this Chapter multilateral and international are meant as synonyms, encompassing the whole range of possibly involved actors.

[2] See also Chap. 2 of this book.

M. Stua, *From the Paris Agreement to a Low-Carbon Bretton Woods*,
DOI 10.1007/978-3-319-54699-5_3

climate change initiatives from an international forum to bilateral and plurilateral regimes[3] would offer more flexibility, thereby creating space for achieving 'more' in smaller groups (Das 2015). Despite negative connotations initially associated with the term 'climate clubs', because they were thought to refer to lobbies and interest groups (Legget 1994), the concept they represent soon gained reputation as a reliable alternative to the international regime of the United Nations Framework Convention on Climate Change—UNFCCC (Sabel and Victor 2015).

Nevertheless, the signing of the Paris Agreement (PA) has been interpreted as an outstanding victory for the international model. Amongst other things, this model ultimately enabled most of the problems raised by the plurilateral[4] frameworks to be overcome. The latter therefore appear to have lost their significance as reliable alternatives to the international climate negotiation system. Yet, the complexity and uncertainty surrounding the implementation of the PA provides a window of opportunity for the development of climate clubs that, acting as complementary tools, can support the effective enhancement and implementation of the PA.

The vast climate club literature has not only focused on negotiating alternatives to multilateralism. A majority of scholars have in fact concentrated their efforts on reimagining plurilateral frameworks as reliable tools that can support the implementation of carbon[5] mitigation strategies within an international climate regime. It is this understanding of plurilateral frameworks, defined by Weischer et al. (2012) as a 'transformational club', which forms the basis of this chapter.

After a brief introduction of the general characteristics referring to climate clubs (Sect. 3.2), this Chapter moves to an in-depth analysis of the rationale (Sect. 3.3), functioning (Sect. 3.4) and enforcement (Sect. 3.5) of transformational clubs. By analysing the various elements that constitute the 'transformational club' framework and by describing its different and sometimes diverging features as well as its dynamic relationship with the international climate negotiation system (Sect. 3.6), this Chapter aims at identifying key elements that may effectively contribute to the foundation of a climate regime capable of meeting the overall objectives of the PA. Summarising the identified elements into a single, experimental governance model defined as the Mitigation Alliance (MA), finally, the concluding Sect. 3.7 suggests a pathway to reconcile the international climate policy regime originated by the PA with the plurilateral perspective offered by the climate clubs.

[3]Plurilateral is here meant as regimes smaller than the international one but encompassing more than two main actors.

[4]From now onwards the term 'plurilateral' is used as synonym of 'minilateral', as bilateral initiatives are not object of discussion of the book.

[5]Carbon Dioxide (CO_2) is the most common Greenhouse Gas (GHG) and is often used as a reference unit to express levels of GHG emissions (expressed in CO_2 equivalent, or CO_2eq). In the context of this Chapter and, more in general, of this book the term carbon, expressing CO_2, and the term GHG are synonyms, hence carbon is used as single term to express GHG.

3.2 The Climate Clubs: Typologies and Interpretations

Climate clubs are usually interpreted as sets of countries that are willing to do 'more', as 'coalitions of the willing' (Das 2015), or again as forums for enthusiastic countries to 'do the deals' (Victor 2015). From this perspective the clubs can be identified as frameworks that facilitate the discussion and promotion of carbon emission mitigation (or simply mitigation), or any sub-issue directly relevant to climate change mitigation (Weischer et al. 2012). Based upon these assumptions, three main typologies of clubs, all characterised by voluntary participation principles, can be envisaged: (a) discussion initiatives, serving as platforms to discuss specific climate-related issues; (b) bargaining forums, aimed at negotiating general climate-related deals; (c) transformational clubs, addressing the adoption of common rules to mitigate carbon emission.

Usually neglected by the clubs' theory, which has mainly concentrated on bargaining forums and transformational clubs, discussion initiatives represent the vast majority of currently existing climate clubs. In their analysis identifying 17 existing climate clubs, Weischer et al. (2012) indirectly emphasise how all of them either belong to the discussion initiatives framework,[6] or focus on narrowly defined subsectors.[7] Sometimes included in the concept of 'pseudo-clubs' (Green 2015), these initiatives produce no direct climate outcome, yet may serve as platforms facilitating the engagement of countries and other entities towards specific climate-related issues and even more ambitious deals, therefore serving as stimulus to the establishment of proper bargaining forums and transformational clubs.

Plurilateral bargaining forums were originally meant to address climate change in the absence of strong international regimes. Naím (2009) for instance regards plurilateral forums as smarter and more targeted alternatives to the obsession with multilateralism. The majority of those advocating bargaining forums address not only the size of the forums, but also the relevance of their participants, as key elements for their effective implementation. Several authors have regarded these forums as clubs comprising a limited number of key players tasked with finding global solutions to climate change (Giddens 2009; Antholis and Talbott 2010). The 'qualitative' character of the forums' members often led to harsh discussions about their real objectives, with several developing countries considering these initiatives as a means for developed countries to avoid responsibility (Hoffmann 2011). Moreover, some commentators have stressed how a multiplicity of plurilateral initiatives instead of a single international strategy may lead to unsatisfactory, if not detrimental, outcomes, due to fragmentation and possible conflicts in law (Das 2015; Victor 2015).

The final outcome of the PA, hailed by most as a clear victory for multilateralism and international climate governance (Christoff 2016; Savaresi 2016; Voigt 2016),

[6] Such as: the G8, the G20, the Carbon Sequestration Leadership Forum, or the Major Economies Forum on Energy and Climate.

[7] Such as: the LEDS Global Partnership, the REDD+ Partnership, or the Global Methane Initiative.

dramatically reduced the significance of plurilateral bargaining forums within the climate change debate. Having lost their role as 'alternatives', the bargaining forums now need to review their objectives in light of the PA's outcomes and its still unresolved issues, in particular those concerning its effective implementation. Such issues still remaining to be solved within the PA regime may offer space for a future revamp of the minilateral bargaining regimes. Clemençon (2016), for instance, stresses how the policy compromise that led to the PA implied the abandonment of some key implementation features which had traditionally formed a core element of climate negotiations, such as the *a priori* definition of binding emission targets and corresponding timetables for signatory parties (so-called sharing), therefore leaving *de facto* a void that still requires to be filled.

In contrast to discussion initiatives and bargaining forums, the promoters of the third typology of clubs, commonly known as transformational clubs, have generally defined them as a complementary tool aimed at enhancing international governance policies (Weischer et al. 2012). Designed to overcome some lock-ins inherent in the international nature of the climate change debate, the 'transformational club' framework may therefore be of greater significance within the PA compared to other climate clubs. For this reason, the key objective of this chapter is the analysis of specific literature on transformational clubs.

3.3 Rationale and Objectives of Transformational Clubs

The rationale behind embracing transformational climate clubs is based on the expectation that such structures may offer greater flexibility and may enable processes leading to outcomes that would be much harder to achieve in larger international forums (Victor 2015). The smaller size of transformational climate clubs may provide an effective space in which greater efforts can be made (Das 2015). The rationale not only concerns the size of groups, but also aims to solve the key problems arising in the implementation of shared efforts to mitigate global carbon emission. More specifically, the club model aims to solve the so-called 'free-riding' problem, which is intrinsic to climate change.

The free-riding issue in climate change is of particular significance because of the inevitable public good that arises out of mitigation efforts (Falkner 2016). Free riding occurs when a party receives the benefits of a public good without contributing to the costs. Hence, as the accumulation of carbon emission happens at atmospheric level with no geographical discrimination, any mitigation action undertaken by one party produces benefits to all other parties alike, regardless of their commitment to undertake similar actions.

Ideally, the free-riding problem may be overcome through the adoption of an international climate regime requiring all parties to act. In this sense, despite its principles that may point to such collective action,[8] the PA did not manage to

[8] See Articles 2 and 4 of the Paris Agreement.

guarantee the required enforcement, being struck by the Westphalian dilemma (Nordhaus 2015), which requires countries to voluntarily consent to join and implement international agreements. As a consequence the bottom-up, voluntary pledges system, of (Intended) Nationally Determined Contributions ((I)NDCs), originally triggered by the 'Lima Call for Action',[9] was formalised by the PA as its means of implementation under the new denomination of Nationally Determined Contributions (NDCs).[10] Based on the voluntariness and sovereignty of the parties in determining and implementing their pledges, the bottom-up system represented by the (I)NDC has not satisfied the enforcement criteria required to overcome the free-riding problem (see below Sect. 3.6). By adopting the (I)NDC model, the PA does not account for the fact that many nations have persistent notions of national interest that render climate protection a low priority.

In contrast, the ambitious 2/1.5 °C objective of the PA[11] appears to imply that participating parties need to design and develop much more ambitious measures for mitigation. In the logic of international cooperation that dominates the aims of the PA, similar measures can be enforced only through the establishment of a mutual-trust system ensuring that efforts made by one party would equate to efforts of other parties (Victor 2015), with the implication that a system must be set up that is able to overcome any free-riding risks.

Transforming mitigation benefits from a public into a private or quasi-private good has been put forward as the only solution to the free-riding issue (Stewart et al. 2013; Falkner 2016) in the absence of more stringent international norms for the enforcement of ambitious mitigation objectives. Clubs literature provides an appropriate framework to test and implement this model. According to the general 'Economic Theory of Clubs' (Buchanan 1965), the core rationale for a club's institution is the provision of private or excludable benefits that are unachievable through other means (Buchanan 1965). Clubs represent a particular type of cooperation game, which aims to produce club benefits that accrue solely or primarily to those that participate in it (Stewart et al. 2013). In principle, clubs find their significance in guaranteeing to their members access to an excludable form of private good. Originally focused on benefits of a financial nature, the theory has evolved towards a variety of interpretations, which have extended the meaning of 'private good' over time.

Transformational clubs find their origins in the so-called Experimentalist Governance (XG). Based upon cooperative game theory in International Environmental Agreements (IEA) studies (Chander and Tulkens 1997), XG aims at providing fertile grounds and rationale for policy experimentations going far beyond the traditional and mainstream pathways. Its application should provide a transformational club with the ability to guarantee to its members effective benefits to stimulate their participation, which generally implies side effects that may include loss of sovereignty on specific climate-related issues. The transformational club as

[9] See Chap. 2 of this book.

[10] See Article 4 of the Paris Agreement.

[11] See Article 2 of the Paris Agreement.

envisaged in this book is therefore an agreement by participating countries to undertake harmonised mitigation (Nordhaus 2015). Creating arrangements that distribute the s of mitigation cost between a determined club of countries, while establishing excludable benefits from their cooperation, represents the strategy for the development of the envisaged structure. The strategy therefore implies that its members perceive mitigation efforts as (part of) the excludable benefits arising out of their affiliation and as unattainable outside the club's domain.

In the literature on clubs, it is generally proposed that this strategy be achieved through a combination of 'carrots' and 'sticks' (Nordhaus 2015; Victor 2015), where 'carrots' represent the benefits for club members and 'sticks' eventually incorporate forms of penalty for non-members and/or for members infringing the club's norms. Additionally, to achieve their promise, the clubs should be designed with rules and features that fit the social, economic and policy circumstances in which they operate (Potoski and Prakash 2009).

An analysis of the structure and functioning of possible forms of climate clubs is required to identify elements that may simultaneously meet the clubs' requirements and their adherence to the current circumstances concerning climate change politics. If these elements are to become ideal pillars for the clubs' structure, some key questions must be solved, including: (a) which models can guarantee that mitigation efforts are transformed from public to private good; (b) which elements can effectively provide tangible benefits to stimulate parties to join the club; (c) and which norms and rules must be established to guarantee the reliability, transparency and effectiveness of the club. Once again, the vast literature on climate clubs offers a variety of suggestions concerning the listed questions, including some contrasting, if not contradictory, views. An analysis of these views in the following sections includes an attempt to understand their rationale and to identify recurring key elements that may effectively serve to promote the design and implementation of an effective transformational club.

3.4 Structure and Functioning of Transformational Clubs

According to Weischer et al. (2012), transformational clubs must embody the following four characteristics to be fully and effectively operational: *speed*, *ambition*, *participation* and *equity*. While *speed* pertains to the idea that a limited group of actors may more easily reach deals and therefore belongs to both the 'bargaining forums' and the transformational clubs, the remaining three elements are of crucial importance in defining the ideal transformational club structure. *Ambition* is perhaps the most important of these elements and it is more achievable in a limited but cohesive group that can more easily agree upon and take charge of relevant 'conditional commitments' (Victor 2015). Conditional commitments represent the key logic for international cooperation, in which the actions of one entity depend on those of another. While the parties' autonomy in defining their own pledges, as established by the PA, clashes with this logic, the introduction of a club system with clear roles in determining its members' quotas would strongly facilitate the establishment of an adequate degree of *ambition*. Based on the analysis of Weischer et al. (2012), *participation* would be linked by the idea that a club would reduce entry barriers for a wide range of stakeholders currently not recognised by the UNFCCC regime. If such a club was to open to entities other than countries, it might enable the development of mitigation strategies tailored towards the requirements of less powerful entities, such as local institutions, cities, firms or small countries. Like *ambition*, *equity* represents one of the two pillars for the Common But Differentiated Responsibilities and Respective Capabilities (CBDR-RC) as recognised by the PA. It calls for the club's members to take into account their differences in determining the distribution of the defined mitigation burden.[12]

Speed, *ambition*, *participation* and *equity* may well represent the characteristics of the ideal transformational club. Nonetheless, several aspects still need to be resolved to enable a shift from a reliable theorisation to an effective set-up of a similar system, insofar as it would be necessary to verify whether a club model could effectively add more value than what might be internationally achievable (Das 2015). In particular, it is vital to understand how an inherent public good such as mitigation might be embedded within a structure that by its nature requires the definition of excludable private goods. The literature on climate clubs emphasises three possible options: an indirect system, where benefits for club members are not climate-related and mitigation happens as a side-effect (Stewart et al. 2013; Orsato et al. 2015); a direct system where mitigation is themself transformed into a private good (Prakash and Potoski 2007; van't Veld and Kotchen 2011); or a hybrid system which combines elements from the other two options (Weischer et al. 2012; Falkner 2016).

At the core of the first option (the indirect system) are strategies and actions aiming for objectives to be achieved through behavioural changes by club members that yield mitigation as a co-benefit (Stewart et al. 2013). The envisaged cooperative regimes are based on the provision of excludable economic, social, or other non-climate benefits for club members (governments, subnational jurisdictions, civil society organisations, firms and consumers), who act out of self-interest and produce net mitigation as a co-benefit. While Orsato et al. (2015) believe these benefits may also include bargaining advantages, such as influencing the corresponding regulatory system, the clubs literature traditionally identifies finance, trade, investment or technology as domains that may offer the benefits required to convert the perceived pay-out from mitigation action from a public good into a private good (Falkner 2016). According to this view the adoption of *ad hoc* rules and regulations for the club's structure and functioning, limited in their application to club members and therefore inherently excludable to non-members, would represent the way to guarantee the members' private benefits. For instance, rules to be applied only to the club's members might include technological transfer and innovation, capacity building, as well as preferential finance and trade systems, applicable in a broad range of

[12] See also Chap. 5, for a wider discussion about the significance of the CBDR-RC within the PA and the present proposal.

sectors including *inter alia*: energy, manufacturing industries, chemical industry, agriculture, forestry, urbanisation, transport, waste management or mining.

The second option (the direct system) leans towards the idea that the recognition of a certification process for mitigation among club members could facilitate a direct shift of mitigation outcomes from a public to a private or quasi-private good (Reid and Toffel 2009; Potoski 2015). The rationale behind such an interpretation relates to the idea that certification would allow members to signal their superior performance compared to competitors acting outside the club's framework, thereby providing a competitive advantage. This logic assumes that members' stakeholders (i.e. those who do business with companies who are members) are directly influenced by the environmental behaviour of companies when they make a choice over whom to do business with. This may happen in two ways: through institutional influence or through behavioural influence. Institutional influence happens when the certification has widespread recognition in terms of (national or international) norms, such as the ISO certification systems (Prakash and Potoski 2006). Behavioural influence is instead based on the significance of civil society choices. This hypothesis is based on the idea that conscious consumers and stakeholders in general may be willing to pay price premiums for goods certifying that they have met environmental or other ethical standards. Behavioural influence is nevertheless poisoned by different elements: 'willingness to pay more' is often overestimated while the perceived benefits tend to be associated with sectors as opposed to single firms, thus facilitating 'free-riding' gains.

Certification as an instrument for shifting the perceived payouts from mitigation efforts from the domain of public to private good, is not only limited to recognition of 'labelling' in its efficacy. A second and significant advantage to the certification of mitigation efforts within a club arises from the opportunity to use the corresponding certifications as tools in carbon markets (Keohane et al. 2015). The possibility for certification to be used as a tool for linking Emissions Trading Schemes (ETS) is highly relevant both for the overall debate on climate clubs and for the purpose of this book: its role is therefore analysed in depth in the following section. The use of certification as a means of linking emissions trading schemes serves to introduce the hybrid option, whereby the private or quasi-private value of mitigation derives from a mix of benefits for the members of the club—those that are climate-related and those that are not.

Weischer et al. (2012) in particular emphasise this label/certification aspect while listing the potential benefits gained from the club. Having stated that the benefits—in order to be effective—need to be significant, exclusive to the club members, accrue to all members of the club, and be generated in a way that respects existing international law, Weischer et al. (2012) provide a detailed list of which potential benefits can be obtained through the establishment of a transformational climate club. The list includes, *inter alia*: linking of emissions trading; trade in goods; trade in services; public procurement; harmonisation and mutual recognition of labels, standards and certification; joint projects and investments; improved investment conditions; joint research and development and sharing of intellectual property rights; access to public climate finance; exemption from Border Carbon Adjustment

(BCA); and agreement on the acceptable level and form of support for green industries.[13]

The rationale behind the significant list of benefits introduced by Weischer et al. (2012) is that items on the list can be excludable and therefore not applicable to entities outside the club's regime. With the benefits representing the 'carrot' of the climate club strategy, the withholding of these benefits, as well as the threat of possible exclusion, should represent its main 'stick' (Das 2015; Nordhaus 2015; Victor 2015). In fact, several of the listed benefits, such as those to do with trade and exemption from BCA, would imply the introduction of sanctions, both against the club's non-members and members who do not accomplish the duties—in the form of assigned shares of mitigation burden, or quotas[14]—agreed by the club. Similar actions are required to enhance the credibility and significance of the club and to limit, if not avoid, the risks of carbon leakage (Keohane et al. 2015; van den Bergh 2016).[15] Nevertheless, the legal aspects concerning the adoption of 'stick' norms to support the club have been widely debated and no clear answer about their feasibility has been given to date. Sanctions represent one of the possible systems to guarantee enforcement of the club's objectives. A detailed analysis of these systems aimed at identifying their respective advantages and disadvantages is provided in Chap. 8 of this book.

3.5 Models to Enforce the Club's Objectives

According to several scholars, a combination of 'stick' and 'carrot' strategies is essential to guarantee the successful implementation of environmental clubs more broadly (Barrett 2003) as well as climate clubs in particular (Sabel and Victor 2015). Therefore, the two models which this book identifies as tools for an effective transformational climate club, the sanctions (i.e. 'stick') and the linking of carbon markets (i.e. 'carrot'), should function as complementary rather than alternative to each other. A hybrid model combining them would therefore represent the ideal solution. Nevertheless, climate club scholars tend to be divided on the subject, with some giving more weight to a sanction-related system (Nordhaus 2015) and others emphasising the potentials of linking carbon markets (Keohane et al. 2015).

Those who advocate the prevalence of a 'stick' mechanism to enforce the club system identify it as a tool able to guarantee more stable coalitions (Nordhaus 2015). Penalties for non-members should, according to this view, produce a double

[13] For a detailed analysis of the listed benefits, see Weischer et al. 2012, pp. 188–191.

[14] Quota is used in this Chapter and more in general in this book as synonym of 'assigned share of mitigation burden'.

[15] Carbon leakage represents a significant uncertainty in climate clubs literature. It serves to state that carbon intensive industries will outsource their productions to areas external to the club and therefore not subject to the club's mitigation burden. Carbon leakage is further discussed in Chaps. 5 and 8 of this book.

effect. On the one hand, they should increase the 'willingness to participate', stimulated by the interest of potential members to avoid incurring the sanctions defined by the club. On the other hand, this model should reduce the risk of withdrawal by offering to current members a clear *ex ante* indication of the costs they would incur by abandoning the club. To be effective, the 'stick' should include mechanisms such as trade sanctions and border tariff adjustments[16] (Sabel and Victor 2015), with particular emphasis paid to the levy of 'carbon equalization tariffs' on carbon-intensive goods imported from non-member stakeholders (van den Bergh 2016).

The main drawback to this system is its legal compatibility with existing international agreements and norms, with special reference to those regulating the World Trade Organisation (WTO) system. While some studies have shown how existing trade law allows the use of similar measures (Cosbey and Mavroidis 2014; Young 2014), other studies on international law are ambiguous on the subject, and questions remain about their legitimacy (Charnovitz 2014). Uncertainty on the subject has also been raised by Das (2015), who underlines how similar measures, when judged strictly in light of the standard public policy discourse, may be interpreted as standing in conflict with the existing international legal framework. Following the establishment of the PA, which did not expressly include norms concerning potential conflicts with other international legislation like the WTO,[17] the debate on the possibility of adopting trade sanctions and border adjustments as mechanisms to enforce a climate club appears still open and far from a definitive solution.

In contrast to the 'stick' system, the use of the climate club model to facilitate the linking of different carbon markets, and/or to promote the establishment of a new single market for club members, appears less problematic in legal terms and still appeals to many scholars, within and outside the specific literature on climate clubs. Two main approaches[18] that link ETS have emerged over the years (Ranson and Stavins 2015). The first is based on bottom-up, direct agreements between the different ETS, while the second is organised according to a top-down, indirect structure, where a single mechanism is adopted to simultaneously link all the different ETS. This second approach represents the model that prevailed during the Kyoto Protocol (KP), leading to the establishment of the Clean Development Mechanism (CDM). An in-depth analysis of the two models is required to identify which advantages and disadvantages they might have within a transformational club, as well as which of the two would be more compatible with the current climate policy regime.

The first, bottom up, direct model is based on the idea of 'recognising' different carbon markets. In principle this recognition can consist of two types: one-way (unilateral) recognition, where only one market recognises another market's mitigation certificates; or two-way (bilateral, plurilateral and international) recognition, where certificates are reciprocally recognised between markets (Jaffe et al. 2009).

[16] Border tariffs adjustments concerning climate-related trade are commonly defined BCA and in such form are generally used in this book.

[17] See also Chap. 8 of this book.

[18] In the context of this chapter, and more in general of this book, the term 'approach' refers to any form of mitigation exchange.

With its members inherently enjoying equal rights, a transformational club based on the direct model falls into the second category of two-way recognition. The most significant example of such an approach within the climate clubs literature is the 'club of carbon markets' (CCM) proposal made by Keohane et al. (2015). Based on the principle that in bottom-up climate architecture, individual jurisdictions establish their own domestic climate policies, including ETS programmes, the CCM proposal aims at promoting deep mitigation ambitions by establishing a linking model for different ETS. Therefore, at the core of the CCM proposal is the mutual recognition of ETS-originated units among members.[19]

To promote the tradability of emission units between the ETS, the CCM would first establish a list of minimum criteria for club membership and therefore for the club's establishment. These would include: (a) the definition of emission targets/emission budgets and corresponding timetables; (b) the use of emissions trading to meet the cap represented by the targets; (c) a jurisdiction to ensure that members comply with the assigned targets; (d) clear, sufficient penalties for noncompliance and institutional capacity to enforce the cap; (e) a commitment to transparent Monitoring, Reporting, and Verification (MRV) for production and exchange of emission reductions. To meet the listed criteria and to recognise the linking model, the CCM would require policies to: (a) establish harmonised or reciprocal standards for monitoring and reporting emissions accounting, transparency and environmental integrity; (b) create a shared market infrastructure to support the mutual recognition of ETS-originated mitigation units; (c) share experience and cooperate in building institutional capacity; (d) work jointly to further the ability of these markets to promote domestic and cross-border investment in low carbon technologies.

At a first glance, the model represented by the CCM proposal appears to be in line with the international climate policy regime of the PA. Generally recognised as a clear victory for the bottom-up architecture for climate action (Savaresi 2016; Voigt 2016), the PA identifies in the (I)NDC the ideal means of implementation for reaching its objectives (see above, Sect. 3.3). Several (I)NDCs already include their own existing or programmed ETS as part of their climate policies and some of them contain expressions of interest towards the implementation of ETS linking tools. Nevertheless, the complexity of the harmonised/reciprocal standards required to set up a direct model for linking ETS, coupled with uncertainties concerning the (I) NDC approach which have already been highlighted by climate policy scholars (Peters et al. 2015; du Pont et al. 2016; Rogelj et al. 2016), still leave windows of opportunity for the implementation of indirect linking models.

Usually interpreted as a top-down approach, the indirect linking of ETS implies the adoption of a unity mechanism, which is able to compare systems of allowances used by the different club members' ETS. In doing so mechanism enables the certified units to freely circulate between various members of the club, regardless of their origin.[20] The most typical application of such an indirect system is represented

---

[19] A more complete analysis of the direct and indirect approaches to markets' reciprocal recognition is offered in Chap. 7 of this book.

[20] See Chap. 7 of this book.

by the KP's CDM. Originally adopted as a carbon offsetting mechanism, the CDM acted as a link between many of the world's cap-and-trade systems, including the ETS of the European Union (EU), Japan, Switzerland and New Zealand (Ranson and Stavins 2015). Despite the uneven results of the CDM application during the KP's first commitment period, the adoption of a CDM-like model for the implementation of an effective climate club is consistent with the aims and objectives of the PA.[21] To understand the degree of consistency, it is first of all necessary to analyse in depth the reciprocal relationship between the literature on climate clubs and the international climate policy regime that emerged after Paris.

3.6 Transformational Clubs and the Paris Agreement Regime

As mentioned above, climate clubs as alternative climate governance models appear to have lost their significance with the establishment of a new and international/multilateral climate regime in Paris. Moreover the new model to mitigation, as included in the PA, challenges the burden sharing principle, which traditionally represents one of the key pillars supporting the climate club rationale (Clemençon 2016). While the overall structure of the Agreement can be seen as a balance between top-down and bottom-up systems (Maljean-Dubois 2016), the latter widely prevailed in defining its mitigation strategies.

The pledge and review (I)NDC, which provides parties with full independence in designing their mitigation policies, represents a turning point whose influence heavily affects the rationale of climate clubs. With their theoretical significance originally established within the wider literature concerning the institution of bottom-up climate regimes (Aldy et al. 2007; Carraro and Egenhofer 2007), climate clubs have now gained a *quasi* top-down appearance compared to the new (I)NDC logic (Nordhaus 2015). Nevertheless, while facing big threats, climate clubs have the potential to also find new opportunities within the PA, if they are correctly framed in the new regime. Reassessing the climate clubs literature and transformational clubs in light of the PA therefore serves both to help understand the impact of the PA on the clubs' regimes and identify the role that clubs may still play in the new climate regime.

The significance of the renewed relationship between climate clubs and the international climate policy regime of the PA lies in the complementarity between the two. For instance, the debate on whether any form of climate club could be hosted under the regime of the UNFCCC has dominated the literature over the years (Das 2015). Several academics in the field have affirmed that a form of complementarity with the international regime is required to guarantee the effective functioning of any form of climate club, consequently implying the possibility for such clubs to be

[21] See Chap. 6 of this book.

hosted under a international framework. Weischer et al. (2012), for example, affirm that an international forum is required to coordinate action between a larger set of countries, while Falkner (2016) envisages a formal link between the club, or clubs, and the international regime as a way of reducing the risk of fragmentation and destabilisation that may arise out of the formation of these clubs.

However, Weischer et al. (2012) also envisage a scenario whereby a smaller group of countries moves ahead in parallel to the UNFCCC, bringing ambition back into the UNFCCC over time. To these authors, the 'bringing back' process may mean either that the club enables parties to take on more ambitious commitments in the UNFCCC, or that club members may negotiate, recognise and monitor commitments or actions within the UNFCCC. The core driver for a strong influence of the UNFCCC regime over the actions and strategies envisaged by the club would be the conferral of legitimacy through the UNFCCC's official recognition of the club's mitigation efforts (Das 2015).

The club's legitimacy is mainly based on the club being able to guarantee an effective and credible measurement, reporting and verification (MRV) system for the actions and strategies undertaken within its regime (Falkner 2016). A reliable MRV system would indeed guarantee a variety of benefits for both the club and the international climate policy regime. It may further reduce free-riding risks (Orsato et al. 2015), support the environmental integrity of actions taken by the club, therefore promoting its transparency (Das 2015), and ultimately facilitate the alignment of club targets with the objectives of the international climate policy regime (Stewart et al. 2013).

While some scholars believe that the club regime possesses more innovative capacity for MRV compared to the international regime (Weischer et al. 2012), the majority believe that incorporating the club's MRV procedures in the wider international jurisdiction of the UNFCCC may substantially support the legitimacy and reliability of the actions undertaken within the club framework (Stewart et al. 2013; Das 2015; Potoski 2015). Ultimately, linking the club regime to the UNFCCC process in a way that respects the need for multilateralism in addressing the global character and consequences of climate change may help enforce the club's legitimacy (Stewart et al. 2013). The adoption of a UNFCCC-established MRV system would represent an ideal linking tool for this purpose.

A final but fundamental possible link between a transformational climate club and the newly established international climate policy regime is the reciprocal influence the two systems could have when it comes to raising mitigation ambitions in line with the PA's objectives. Such influence can be seen through the lenses of two opposite perspectives. On the one hand, promoters of the CCM envisage in the currently established bottom-up strategy identified in the voluntary pledges of the (I) NDC the ideal framework for the implementation of the CCM system (Keohane et al. 2015). For instance the CCM capability to directly link different ETS systems may represent a key tool for those parties that included the adoption of international carbon trade among the instruments listed in their (I)NDCs for reaching their self-determined pledges.

On the other hand, an increasing number of scholars have expressed concern about the effectiveness of the (I)NDC model in raising mitigation ambitions (Caparrós 2016; Clemençon 2016; Morgan 2016), while others have already stressed the need for additional and complementary strategies to effectively enhance these ambitions (Barrett and Dannenberg 2016; Rogelj et al. 2016). Given that the uncertainty concerning the effectiveness of the (I)NDC model appears unsolvable by the CCM, climate clubs scholars are currently stimulated to search for alternative solutions. Sabel and Victor (2015) recognise the weakness of the (I)NDC-based pledges as tools for achieving the ambitious goals of the PA emphasising that the highly discretionary model that characterises the design and implementation of the (I)NDCs offers no guarantee of their efficacy. In contrast, they promote the idea of a transformational club based on the original meaning of XG. From their perspective, an experimental governance model structured on a club's ability to guarantee effective benefits to its members could provide those enforcements lacking in the (I)NDC model. Using the experimental governance application within the PA could indeed lead to a stronger system of pledging through the establishment of quantified mitigation targets shared by all club members, through the design of stronger and more credible penalty defaults for non-compliant members, and finally through the definition of collaborative compliance systems. An innovative model of this kind, which is seemingly dismissed by the overall bottom-up system adopted by the PA, actually does find support within Article 6 of the PA (A6PA) that represents the foundation for the proposal promoted by this book. Based upon the concept of voluntary participation for higher ambitions in mitigation compared to the (I)NDCs, A6PA can be interpreted as the ideal trigger for the implementation of innovative and experimental governance regimes like the one introduced in this book.[22]

3.7 Conclusions: Towards a Mitigation Alliance

After briefly surveying the vast literature on climate clubs and the multitude of models and models for their design and development, the main advantages of bringing these together as a means of boosting the international climate governance regime should now be clear. The overall importance of climate clubs lies in their capacity to promote higher ambition in mitigation by linking the latter to private goods paid out to club members. As Das (2015, p. 24) has emphasised, 'to the extent [that] the club benefits succeed in generating climatic benefits, say greater climate mitigation, those would of course be public goods.' Climate clubs pose such great potential for climate governance because they reward effective climate action with private goods for participating club members, while at the same time safeguarding the inherent significance of mitigation as a public good.

In the context of this book, the importance of climate clubs lies in their following capabilities:

[22] See especially Chap. 5 of this book.

- Encouragement of higher ambition in mitigation; by
- Transforming *carbon* into a private good; while
- Safeguarding the inherent significance of *carbon* as a public good.

Although several proposals on climate clubs appear to be overtaken by the achievement of an international agreement on climate change (the PA), some can still be interpreted as adhering to the newly established climate policy regime. Despite the CCM perspective appearing to be more in line with the PA and easily linked to the bottom-up system founded on the (I)NDCs, it is unable to ensure that the primary objective for both the PA and climate clubs is reached, namely the promotion of high mitigation ambitions. The absence of a reliable and enforceable regulatory system (Daskalakis and Markellos 2009; Cao and Xu 2016) and the lack of quantified abatement targets (Blyth et al. 2009; Kim 2015), both of which remain unsolved by the adoption of the (I)NDCs, represent the biggest obstacles to any form of club in reaching its primary objective.

With the limits posed by the bottom-up, pledge-based system adopted by the PA directly affecting the CCM system's possible effectiveness, the adoption of an experimental governance model appears to be the main, if not the only, pathway for the implementation of an effective club able to fulfil its primary objective. As further explained in Chap. 4 of this book, the legal provisions included in A6PA can provide the necessary regulatory framework for the correct implementation of an experimental governance model in the shape of a transformational club, composed of entities that voluntarily decide to participate in the 'experimentation', whose final aim would be the institution of a 'low carbon union' to effectively provide mitigation in line with the PA's objectives. This ideal model of an effective transformational club is hereafter called the Mitigation Alliance (MA). Through an in-depth analysis of the structure and significance of A6PA, and verification of its convergence with the transformational clubs conceptualisation, the next chapter provides the fundamental pillars for the establishment of an effective Mitigation Alliance (MA).

References

Aldy, J. E., Stavins, R. N., Frankel, J. A., & Summers, L. H. (2007). *Architectures for agreement: Addressing global climate change in the post-Kyoto world*. Cambridge: Cambridge University Press.

Antholis, W., & Talbott, S. (2010). *Fast forward: Ethics and politics in the age of global warming*. Washington, DC: Brookings Institution.

Barrett, S. (2003). *Environment and statecraft: The strategy of environmental treaty making*. New York: Oxford University Press.

Barrett, S., & Dannenberg, A. (2016). An experimental investigation into 'pledge and review' in climate negotiations. *Climatic Change, 138*, 339–351.

Blyth, W., Bunn, D., Kettunen, J., & Wilson, T. (2009). Policy interactions, risk and price formation in carbon markets. *Energy Policy, 37*, 5192–5207.

Buchanan, J. (1965). An economic theory of clubs. *Economica, 32*, 1–14.

Cao, G., & Xu, W. (2016). Nonlinear structure analysis of carbon and energy markets with MFDCCA based on maximum overlap wavelet transform. *Physica A, 444*, 505–523.

Caparrós, A. (2016). The Paris agreement as a step backward to gain momentum: Lessons from and for theory. *Revue d'Economie Politique, 126*(3), 347–356.

Carraro, C., & Egenhofer, C. (2007). *Climate and trade policy: Bottom-up approaches towards global agreement*. Cheltenham: Edward Elgar Publishing Limited.

Chander, P., & Tulkens, H. (1997). The core of an economy with multilateral environmental externalities. *International Journal of Game Theory, 26*(3), 379–401.

Charnovitz, S. (2014, September 29). Border tax equalization. Draft paper prepared for the conference on challenges facing the world trade system. Columbia University and Johns Hopkins University.

Christoff, P. (2016). The promissory note: COP 21 and the Paris climate agreement. *Environmental Politics, 25*(5), 765–787.

Clemençon, R. (2016). The two sides of the Paris climate agreement: Dismal failure or historic breakthrough? *Journal of Environment and Development, 25*(1), 3–24.

Cosbey, A., & Mavroidis, P. C. (2014). Heavy fuel: Trade and environment in the GATT/WTO case law. *RECIEL, 23*, 288–301.

Das, K. (2015). Climate clubs: Carrots, sticks and more. *Economic & Political Weekly, 50*(34), 24–27.

Daskalakis, G., & Markellos, R. (2009). Are electricity risk premia affected by emission allowance prices? Evidence from the EEX, Nord Pool and Powernext. *Energy Policy, 37*, 2594–2604.

Du Pont, Y. R., Jeffery, M. L., Gütschow, J., Christoff, P., & Meinshausen, M. (2016). National contributions for decarbonizing the world economy in line with the G7 agreement. *Environmental Research Letters, 11*, 054005.

Falkner, R. (2016). A minilateral solution for global climate change? On bargaining efficiency, club benefits and international legitimacy. *Perspectives on Politics, 14*(1), 87–101.

Giddens, A. (2009). *The politics of climate change*. Cambridge: Polity Press.

Green, J. F. (2015). The strength of weakness: Pseudo-clubs in the climate regime. *Climatic Change*. doi:10.1007/s10584-015-1481-4.

Hoffmann, M. J. (2011). *Climate governance at the crossroads: Experimenting with a global response after Kyoto*. New York: Oxford University Press.

Jaffe, J., Ranson, M., & Stavins, R. N. (2009). Linking tradable permit systems: A key element of emerging international climate policy architecture. *Ecology Law Quarterly, 36*, 789–808.

Keohane, N., Petsonk, A., & Hanafi, A. (2015). Toward a club of carbon markets. *Climatic Change*. doi:10.1007/s10584-015-1506-z.

Kim, M.-K. (2015). Dynamic interactions between carbon and energy prices in the US regional greenhouse gas initiative. *International Journal of Energy Economics and Policy, 5*(2), 494–501.

Legget, J. (1994). Trends in the governmental and business response to climate change: Problems for the carbon club, opportunities for the renewable energy sector. *Renewable Energy, 5*(2), 1311–1313.

Maljean-Dubois, S. (2016). The Paris agreement: A new step in the gradual evolution of differential treatment in the climate regime? *RECIEL, 25*(2), 151–160.

Morgan, J. (2016). Paris COP 21: Power that speaks the truth? *Globalizations*. doi:10.1080/14747731.2016.1163863.

Naím, M. (2009). Minilateralism: The magic number to get real international action. *Foreign Policy, 173*, 135–136.

Nordhaus, W. (2015). Climate clubs: Overcoming free-riding in international climate policy. *American Economic Review, 105*(4), 1339–1370.

Orsato, R. J., de Campos, J. G. F., Barakat, S. R., Nicolletti, M., & Monzoni, M. (2015). Why join a carbon club? A study of the banks participating in the Brazilian "business for climate platform". *Journal of Cleaner Production, 96*, 387–396.

Peters, G. P., Andrew, R. M., Solomon, S., & Friedlingstein, P. (2015). Measuring a fair and ambitious climate agreement using cumulative emissions. *Environmental Research Letters, 10*, 105004.

Potoski, M. (2015). Green clubs in building block climate change regimes. *Climatic Change*. doi:10.1007/s10584-015-1517-9.

Potoski, M., & Prakash, A. (2009). *Voluntary programs: A club theory approach*. Cambridge, MA: MIT Press.

Prakash, A., & Potoski, M. (2006). Racing to the bottom? Globalization, environmental governance, and ISO 14001. *American Journal of Political Sciences, 50*, 347–361.

Prakash, A., & Potoski, M. (2007). Collective action through voluntary environmental programs: A club theory perspective. *Policy Studies Journal, 35*(4), 773–792.

Ranson, M., & Stavins, R. N. (2015). Linkage of greenhouse gas emissions trading systems: Learning from experience. *Climate Policy*. doi:10.1080/14693062.2014.997658.

Reid, E. M., & Toffel, M. W. (2009). Responding to public and private politics: Corporate disclosure of climate change strategies. *Strategic Management Journal, 30*(11), 1157–1178.

Rogelj, J., den Elzen, M., Hohne, N., Fransen, T., Fekete, H., Winkler, H., Schaeffer, R., Sha, F., Riahi, K., & Meinshausen, M. (2016). Paris agreement climate proposals need a boost to keep warming well below 2 °C. *Nature, 354*, 631–639.

Sabel, C. F., & Victor, D. G. (2015). Governing global problems under uncertainty: Making bottom-up climate policy work. *Climatic Change*. doi:10.1007/s10584-015-1507-y.

Savaresi, A. (2016). The Paris agreement: An early assessment. *Environmental Policy and Law, 46*(1), 14–18.

Stewart, R., Oppenheimer, M., & Rudyk, B. (2013). A new strategy for global climate protection. *Climatic Change, 120*, 1–12.

Van den Bergh, J. C. J. M. (2016). Rebound policy in the Paris agreement: Instrument comparison and climate-club revenue offsets. *Climate Policy*. doi:10.1080/14693062.2016.1169499.

Van't Veld, K., & Kotchen, M. (2011). Green clubs. *Journal of Environmental Economics and Management, 62*, 309–322.

Victor, D. G. (2015, January). The case for climate clubs. *E15 Initiative Thinking Piece*.

Voigt, C. (2016). The compliance and implementation mechanism of the Paris agreement. *RECIEL, 25*(2), 161–173.

Weischer, L., Morgan, J., & Patel, M. (2012). Climate clubs: Can small groups of countries make a big difference in addressing climate change? *RECIEL, 21*(3), 177–192.

Young, M. A. (2014). Trade measures to address environmental concerns in faraway places: Jurisdictional issues. *RECIEL, 23*, 302–317.

References

Chapter 4
Article 6 of the Paris Agreement as Foundation for the Mitigation Alliance

Abstract Although it has been generally hailed as a breakthrough in international climate change policy, several unsolved issues still surround the Paris Agreement (PA). Misalignment between its ambitious objectives (the 2 °C/1.5 °C target) and its designated means of implementation (the Nationally Determined Contributions—NDCs) represents the biggest challenge. Parties are not required to strictly align their NDCs with the PA's 2 °C/1.5 °C objectives as a result of the Common But Differentiated Responsibilities and Respective Capabilities principle (CBDR-RC). Nevertheless, the PA includes a clause, represented by its Article 6 (A6PA), which offers a window of opportunity to overcome this impasse. Limited to parties that voluntarily collaborate for higher mitigation ambition, hence bypassing the possibility of vetoing that affected the history of climate negotiations, A6PA allows for experimentation with new models of climate governance capable of meeting all the PA's objectives. Identifying pathways towards experimental climate governance models requires an in-depth analysis of the A6PA's wording, meaning, and significance, as well as its embedding within experimental governance literature, especially the one surrounding transformational clubs. This chapter concludes with the identification of a skeletal architecture for a new governance model: the Mitigation Alliance (MA).

Keywords Paris Agreement (PA) • Article 6 (A6PA) • Experimental governance • Transformational clubs • Mitigation Alliance (MA)

4.1 Introduction

The Paris Agreement (PA) is the fruit of years of diplomatic action and scientific effort. While attempting to balance the different needs, interests and ideologies of the parties, the PA pays high attention to both life and social sciences. Its final objective and most of its structure are aligned with the 5th report of the Intergovernmental Panel on Climate Change's (IPCC) findings and recommendations (IPCC 2014). Moreover, social, economic and political sciences have contributed substantially to the design of its architecture. Dozens of research centres, schools and institutes from all over the world have, often jointly, elaborated a sound scientific framework that provides the foundation for a new, climate-friendly regime.

M. Stua, *From the Paris Agreement to a Low-Carbon Bretton Woods*,
DOI 10.1007/978-3-319-54699-5_4

It is likely that similar contributions will continue to play an important role throughout the PA's implementation phase.

However, the final outcome of the PA solves only some of those 'obstacles to a global treaty' perceived by Stewart et al. (2013), while some omissions have already become evident (Bodansky 2016). Misalignment between national policies and scientific recommendations are of great concern to the climate change policy community (Clémençon 2016; Van Asselt 2016; van den Bergh 2016). At the current stage of the negotiations, no methodological model has been established to guarantee the implementation of activities for carbon[1] emission mitigation (or simply mitigation) in line with the ambitious PA objectives (Waisman et al. 2016; Van Asselt 2016). The objectives mandated by the PA are also unclear on how effectiveness is to be combined with equity, as implied in the concept of Common But Differentiated Responsibilities and Respective Capabilities (CBDR-RC) (Rajamani 2016; Waisman et al. 2016). When applied to all the parties, the PA does not provide specific guidance; it leaves high degrees of uncertainty, not only regarding the implementation of its objectives, but also in respect of the participation of non-state entities (Van Asselt 2016; Waisman et al. 2016) and the transparency of the overall regime (Rajamani 2016; Waisman et al. 2016). Yet the PA contains an article that enables governance regime experimentation among a limited number of parties (and/or non-state entities authorised by their reference parties), which may provide the basis for aligning ambition and action more closely with scientific recommendations: Article 6 (Caparrós 2016; Marcu 2016; van Asselt 2016).

Article 6 of the PA (A6PA) is considered one of its finest outcomes of the policy compromises. Completed just a few hours before the final approval of the PA text, A6PA is the result of a nuanced set of different, and somehow diverging, interests. As described by Marcu (2016), the Article's approval necessitated taking into account a multitude of factors, such as: (a) its significance in relationship to the environmental integrity of the PA; (b) the multiple linkages with other important parts of the Agreement; (c) the need for a balance between the idea of 'tradable' mitigation in carbon markets and non-market approaches negotiated in Paris; and (d) the ideological opposition to a direct or too explicit inclusion of markets within the final text of the Agreement. Intentionally broad in its interpretation (Marcu 2016), A6PA potentially envisages all types of cooperation for enhancing ambition and progress in climate change combat. Implicitly experimental thanks to the voluntary participation principle, A6PA can be interpreted as a trigger for testing and verifying the effectiveness of new climate governance regimes (Caparrós 2016). A6PA may therefore be of utmost significance.

The aim of this book is to propose a multi-level, hybrid governance model applicable through A6PA. Combining old and new elements in the realms of policy, economy, finance and society, the model comprises few but well defined rules for

[1] Carbon Dioxide (CO_2) is the most common Greenhouse Gas (GHG) and is often used as a reference unit to express levels of GHG emissions (expressed in CO_2 equivalent, or CO_2eq). In the context of this Chapter and, more in general, of this book the term carbon, expressing CO_2, and the term GHG are synonyms, hence carbon is used as single term to express GHG.

Table 4.1 Links between A6PA and other Articles of the Paris agreement

Articles of the Paris agreement	Linked paragraphs of A6PA
Article 2	Paragraphs 1, 2, 4 and 8
Article 4	Paragraphs 1, 2, 3, 4, 5 and 8
Article 7	Paragraphs 1, 6 and 8
Article 9	Paragraphs 2, 4 and 8
Article 10	Paragraph 8
Article 11	Paragraph 8
Article 13	Paragraphs 2 and 5
Article 16	Paragraphs 2, 4, 6 and 7

those entities willing to 'cooperate' under its umbrella. These include, the key aims of the PA, thereby transforming the regime into a tool to promote sustainable development as well as quantifiable mitigation and equity in the new international climate framework.

To understand the architecture of the governance regime, which corresponds to the Mitigation Alliance (MA) introduced in Chap. 3 of this book, it is necessary to analyse A6PA from an experimentalist governance perspective. This chapter first discusses the text of A6PA in detail, analysing its most significant passages and describing their relevance for the implementation of the PA's objectives (Sect. 4.2). Secondly, this chapter pays special attention to the wording of A6PA (Sect. 4.3) to identify its most characteristic features. Thirdly, it focuses on the analogies between A6PA and transformational clubs governed by game theory literature[2] (Sect. 4.4). The findings serve to explain the significance of A6PA for the establishment of the MA and also to guarantee the MA's conformity to the PA. To reaffirm this significance and conformity, this Chapter identifies ten key norms and criteria that facilitate some preliminary considerations on the MA's architecture (Sect. 4.5).

4.2 Unveiling Article 6

As with many other sections of the PA, A6PA was not only the product of diplomacy and compromise but first and foremost the product of science. Understanding the features of the governance models proposed by A6PA first requires the identification of its scientific references. On the one hand, its climate change and sustainability aims—like the entire PA—follow the recommendations of the IPCC. On the other hand, its political and economic structure finds its origins in the climate clubs' literature, as well as in the game theory framework (Weischer et al. 2012; Stewart et al. 2013; Caparrós 2016). The scientific resources are not separate within the text of A6PA, but rather synergistically combined. A correct interpretation of A6PA's

[2] See also Chap. 3 of this book.

peculiar wording, also in comparison to the remaining PA text, reveals a trigger for the establishment of new governance regimes.

A6PA refers directly to nine articles contained in the Agreement. Article 5 (A5PA), in particular, appears bound to A6PA with special reference to the possible interactions between the mechanism of A6PA (Paragraphs 4 to 7) and the sinks-related actions described in A5PA (Marcu 2016, p. 2). While the links between A5PA and A6PA will be further analysed in Chap. 7 of this book, Table 4.1 introduces A6PA's links with other Articles of the PA.

A6PA appears to play a central role in the overall architecture of the PA and its significance can only be understood by reviewing each of its components.

Comprising nine paragraphs, A6PA is one of the most complex parts of the entire PA, thus requiring special attention from those attempting to understand the meaning and significance of the PA. Accurate paraphrasing of its text is required to facilitate comprehension. A6PA is organised into three parts, with the central one (Paragraphs 4 to 7) focusing on the institution of the mechanism intended to stimulate higher mitigation ambition for the entities voluntarily cooperating under the A6PA,[3] anticipated by a three-Paragraph introductory part and followed by two final Paragraphs. The analysis of the text proposed here follows this structure. The text of each Paragraph is reproduced in italic script and paraphrased in brackets, with detailed commentaries at the end of each part.

Once the first analysis of A6PA has been completed, a second analytical step helps identify and understand the key words included in its text, as well as their significance and relevance in defining the basic criteria for the design of coherent governance regimes to be applied through A6PA.

(A6PA, introductory section: Paragraphs 1–3)

Paragraph 1 Parties recognize that some Parties choose to pursue voluntary cooperation in the implementation of their nationally determined contributions to allow for higher ambition in their mitigation and adaptation actions and to promote sustainable development and environmental integrity. (Some parties may voluntarily establish cooperative frameworks to enhance their mitigation and/or adaptation ambitions and promote sustainable development).

Paragraph 2 Parties shall, where engaging on a voluntary basis in cooperative approaches that involve the use of internationally transferred mitigation outcomes towards nationally determined contributions, promote sustainable development and ensure environmental integrity and transparency, including in governance, and shall apply robust accounting to ensure, inter alia, the avoidance of double counting, consistent with guidance adopted by the Conference of the Parties serving as the meeting of the Parties to the Paris Agreement. (Within the framework of voluntary cooperation, parties shall exchange mitigation outcomes and thus guarantee sustainable development promotion, environmental integrity and transparency, consistent with Article 13 and strive to avoid double counting).

[3] See also Chap. 6 of this book.

Paragraph 3 The use of internationally transferred mitigation outcomes to achieve nationally determined contributions under this Agreement shall be voluntary and authorized by participating Parties. (The mitigation outcomes may become the unity measure for voluntary exchanges of contributions authorised by the parties).

A6PA introduces voluntary cooperation as a framework to support parties in enhancing their mitigation/adaptation ambitions and promoting sustainable development. It affirms that any action taken within the voluntary cooperation framework must be consistent with the environmental integrity and transparency principles and that double counting must be avoided. It states that Internationally Transferred Mitigation Outcomes (ITMOs) resulting from voluntary cooperation can count towards Nationally Determined Contributions (NDCs) when authorised by the parties involved, thus confirming the Article's potential value as supplier of mitigation outcomes.

In Paragraphs 1 to 3, A6PA identifies, in the process of voluntary cooperation, a new and experimental instrument to enhance the objectives of the PA, possibly through the adoption of new governance regimes. It states that any mitigation effort under the umbrella of voluntary cooperation and consistent with the mandates of the PA can contribute to NDCs by encouraging more ambitious mitigation (and adaptation) targets. A6PA assigns a fixed mitigation value with a corresponding flexible economic value to mitigation outcomes, regardless of their geographical origin.[4] A6PA defines these outcomes as ITMOs.

(A6PA, central section: Paragraphs 4–7)

Paragraph 4 A mechanism to contribute to the mitigation of greenhouse gas emissions and support sustainable development is hereby established under the authority and guidance of the Conference of the Parties serving as the meeting of the Parties to the Paris Agreement for use by Parties on a voluntary basis. It shall be supervised by a body designated by the Conference of the Parties serving as the meeting of the Parties to the Paris Agreement, and shall aim: (a) To promote the mitigation of greenhouse gas emissions while fostering sustainable development; (b) To incentivize and facilitate participation in the mitigation of greenhouse gas emissions by public and private entities authorized by a Party; (c) To contribute to the reduction of emission levels in the host Party, which will benefit from mitigation activities resulting in emission reductions that can also be used by another Party to fulfil its nationally determined contribution; and (d) To deliver an overall mitigation in global emissions. (A mechanism to facilitate and support mitigation efforts is established. It requires the designation of an international body entitled to supervise the application of the mechanism. The mechanism aims to: promote mitigation and sustainable development; encourage mitigation action by authorised public and private entities (stakeholders); reduce emissions levels of either the host party or the party acquiring the mitigation outcomes; and reduce emissions on aggregate).

---

[4] See Chap. 9 of this book.

Paragraph 5 Emission reductions resulting from the mechanism referred to in paragraph 4 of this Article shall not be used to demonstrate achievement of the host Party's nationally determined contribution if used by another Party to demonstrate achievement of its nationally determined contribution. (Double counting of mitigation outcomes must be avoided).

Paragraph 6 The Conference of the Parties serving as the meeting of the Parties to the Paris Agreement shall ensure that a share of the proceeds from activities under the mechanism referred to in paragraph 4 of this Article is used to cover administrative expenses as well as to assist developing country Parties that are particularly vulnerable to the adverse effects of climate change to meet the costs of adaptation. (A share of the proceeds of mitigation outcomes shall be used to cover administrative expenses and to assist developing and least developed countries in adaptation).

Paragraph 7 The Conference of the Parties serving as the meeting of the Parties to the Paris Agreement shall adopt rules, modalities and procedures for the mechanism referred to in paragraph 4 of this Article at its first session. (The Conference of Parties serving as the meeting of the Parties to the Paris Agreement (CMP) shall establish rules, modalities and procedures for the mechanism).

In its second section (Paragraphs 4–7), A6PA outlines the funding elements for a mechanism to facilitate and accelerate mitigation among parties that decide to voluntarily collaborate at the light of A6PA's Paragraph 1. A mechanism governing the voluntary cooperation in general and, more specifically, the exchange of mitigation outcomes under the scrutiny of an international entity created by the CMP is established in Paragraph 4 of A6PA. This Paragraph also specifies the aims of the mechanism: (a) to promote mitigation and sustainable development; (b) to engage authorised public and private entities (stakeholders) to participate in mitigating actions; (c) to reduce emissions levels of either the host party or the party acquiring the mitigation outcomes; and (d) to reduce emissions on aggregate. By defining minimum criteria for certification and exchange processes, the resulting mechanism shall therefore assign a value to mitigation resulting from voluntary cooperation measures.

While the establishment of most rules, modalities and procedures concerning the functioning of the mechanism awaits future decisions of the CMP, primary importance is given to avoiding double counting. In Paragraph 5, A6PA clearly states that only the final acquirer of the outcome of mitigation effort can claim its value by demonstrating its corresponding mitigation impact, while the original producer shall discount it once the exchange procedure has been completed.

A6PA also affirms that a share of the financial proceeds from outcomes of recognised mitigating efforts shall be used to cover the administrative costs required to maintain the mechanism-related system, thus facilitating self-sustainability. An additional undefined share shall be used to support the costs of adaptation for particularly vulnerable developing countries, thus including an automatic adaptation tool within a mainly mitigation-oriented framework.

(A6PA, final section: Paragraphs 8–9)

Paragraph 8 Parties recognize the importance of integrated, holistic and balanced non-market approaches being available to Parties to assist in the implementation of their nationally determined contributions, in the context of sustainable development and poverty eradication, in a coordinated and effective manner, including through, inter alia, mitigation, adaptation, finance, technology transfer and capacity-building, as appropriate. These approaches shall aim to: (a) Promote mitigation and adaptation ambition; (b) Enhance public and private participation in the implementation of nationally determined contributions; and (c) Enable opportunities for coordination across instruments and relevant institutional arrangements. (Under the framework of voluntary cooperation between parties, non-market, holistic approaches can be integrated within the mechanism or developed in parallel. Their general objectives call for coordinated action encompassing, inter alia, mitigation, adaptation, finance, technology transfer and capacity building. These approaches shall aim to promote mitigation and/or adaptation ambitions, enhance public and private participation in NDC, and stimulate coordination across instruments and institutional arrangements).

Paragraph 9 A framework for non-market approaches to sustainable development is hereby defined to promote the non-market approaches referred to in paragraph 8 of this Article. (The significance of a framework for non-market approaches is stressed).

A6PA concludes by discussing non-market, holistic approaches which, under the framework of voluntary cooperation between parties, can possibly be integrated within the mechanism or developed in parallel. These approaches shall *inter alia* support mitigation, adaptation, finance, technology transfer and capacity building, with the aim to promote mitigation/adaptation ambitions, enhance public and private participation in implementing NDCs, and stimulating coordination across instruments and institutional arrangements. The Article ends with a call to define a framework for non-market approaches to sustainable development. Non-market, holistic approaches to mitigation, adaptation, finance, technology transfer and capacity building are therefore envisaged in Paragraphs 8 and 9. These final paragraphs are of vital importance as they establish that economic sustainability is not the only value to be taken into account when pricing the outcomes of mitigation efforts described in A6PA.

Identifying the significance of A6PA within the PA, as well as analysing the general meaning of its Paragraphs, represents a first step in the definition of its role as a promoter of transformational low carbon pathways. An additional analysis of the wording adopted in key passages of A6PA is required to fully understand its relevance as a trigger for the establishment of experimental governance regimes for the effective implementation of the ambitious PA objectives.

4.3 Analysing the Terminology of Article 6

As for any other part of the Agreement, the terminology adopted in A6PA represents more than the ultimate political compromise between parties, defining instead its own legal significance. Understanding its wording becomes a necessary step in the process of unveiling its full potential and the interrelated opportunities for climate governance experimentation. Section 4.3 identifies, and analyses various key terms adopted in A6PA by taking into account their meaning and objectives. The final result of this terminology analysis represents a complex but solid and homogeneous structure for defining the primary elements of governance models that might eventually be developed through the application of A6PA.

Three terms adopted in A6PA serve to explain and define its relevance as a tool for triggering the establishment of new climate governance regimes: *some parties*, *cooperation* and *voluntary*. The first term is used only once in Paragraph 1 and asserts that, unlike any other part of the PA, A6PA applies only to those parties willing to act under its framework. This term provides A6PAwith the necessary 'freedom' to act as a trigger for experimentation, bypassing any veto risk by parties that do not wish to be involved in any initiative evolving from the A6PA's application. Cited twice (in Paragraphs 1 and 2), *cooperation* represents the second key term to determine the experimental nature of A6PA. It means that any initiative generated by A6PA requires cooperation among stakeholders, thus excluding unilateral actions. The existing link between *cooperation* and *voluntary*, with the former being tied to the latter in both its citations, aims at specifying how any cooperative action developed under A6PA's umbrella cannot be eventually imposed upon parties, requiring instead a clear 'willingness to participate'. This element confirms the underlying 'freedom to participate' concept that characterises the nature of A6PA. It also emphasises the opportunity to interpret A6PA as an experimental governance tool. *Voluntary* is used twice more within the text, in Paragraphs 3 and 4. This use aims at confirming how voluntary participation imbues the whole architecture of A6PA, thus representing a final exclusion guarantee for those parties that do not intend to participate. Once again, this indirectly implies the opportunity to adopt experimental approaches to climate governance for the parties 'willing to participate'.

In analysing the most common terms within the text of A6PA, three topics appear to recur: *mitigation* (cited 10 times), *NDCs* (cited 8 times) and *sustainable development* (cited 6 times). The second and third terms are similarly and evenly distributed across A6PA, being used in the same Paragraphs 5 out of 6 times (Paragraphs 1, 2, 3, 4 and 8). This distribution defines two pillars within the Article: (a) regardless of the nature of cooperation, any voluntary cooperation established under A6PA shall aim at sustainable development; (b) A6PA primarily serves at assisting the NDCs implementation while encouraging an increase in their ambitions. The use of *mitigation* calls for various considerations. Despite being the most common term of A6PA and despite being present in 5 Paragraphs, it registers a unique peculiarity: it appears 5 times in a single Paragraph. This is Paragraph 4, which is entirely focused on the *mechanism* proposed by A6PA. *Mitigation* is used to introduce the *mechanism* and to describe each of its four aims. Its repeated use does not aim only to reaffirm the primary rationale behind the suggested *mechanism*. It also establishes some pillars for the implementation of new governance regimes. The term *higher ambition*, which appears once in Paragraph 1 explaining the objectives of voluntary cooperation, reiterates the priority of A6PA: to accelerate the mitigation processes and promote sustainable development while acknowledging the significance of facilitating adaptation. *Adaptation* appears to play a less significant role in A6PA compared to other sections of the PA, despite four appearances and its inclusion in Paragraph 1. However, *adaptation* is not associated with a defined instrument to facilitate its implementation process and/or the increase in its related ambitions.

Now that the experimental, cooperative and voluntary nature, targets (the NDCs) and areas of intervention (mitigation, sustainable development and adaptation) of A6PA have been identified, this terminology analysis moves towards elements which define most clearly the architecture and functioning of the proposal embedded in A6PA and developed in this book.

The terms *mechanism* and *ITMOs* serve as a first definition of the overall architecture that a governance regime resulting from A6PA requires in order to be consistent with the aims and objectives of A6PA. Appearing four times in Paragraphs 4, 5, 6 and 7, the *mechanism* can be seen as the core innovation of A6PA. It represents a single system to certify any mitigation outcome originated by the application of A6PA, therefore acting as a link between the different approaches to mitigation outcomes' exchange envisaged by the Article.[5] Based on its use in Paragraph 2, the term *ITMOs* may appear at a first sight linked only to bilateral mitigation initiatives involving parties as direct stakeholders. Nevertheless, the second use of *ITMOs*, in Paragraph 3, provides elements for a different and more inclusive interpretation. While stating that its use "shall be voluntary and authorized by participating Parties", Paragraph 3 implies that *ITMOs* are to be used not only by parties but also by authorised stakeholders, thereby overriding its first description. By specifying that parties shall authorise the use of *ITMOs*, Paragraph 3 in fact implies that *ITMOs* are to be used beyond the perimeter of parties' direct sphere of influence. According to this view, it appears more likely that *ITMOs* could represent certificates resulting from the application of the *mechanism*, which could potentially be used in any approach to mitigation outcomes' exchange envisaged by A6PA.[6]

The term *approaches* appears five times in the text of A6PA and serves to outline further the general architecture of the governance models eventually enabled by the framework of A6PA. While reaffirming the cooperative nature of the architecture of A6PA (in Paragraph 2), the term defines the scope of action for the *mechanism* and its ruling system. By explicitly referring to non-market mitigation *approaches* (Paragraphs 8 and 9), as well as to bilateral exchanges of mitigation outcomes between parties (Paragraph 2), the term has been construed by the most recent

[5] For a wider analysis of the significance and role of mitigation outcomes certification on the overall structure of the climate governance proposal addressed by this book see Chap. 6.

[6] See also Chap. 7 of this book.

literature to imply the inclusion of market-oriented *approaches* within the system proposed by A6PA (Marcu 2016; van Asselt 2016). The application range for A6PA and its rules appears holistic, including virtually any possible approach to mitigation outcomes' exchange and merging them under the framework of a single accounting and verification system represented by the *mechanism* introduced in Paragraphs 4 to 7.[7]

Based on this understanding of the main architecture of A6PA, it is now possible to focus on its functioning, which is primarily determined by the A6PA own aims. These include, but are not limited to, above-mentioned mitigation, sustainable development and adaptation as well as the linked concepts of *environmental integrity* (in both Paragraphs 1 and 2) and *transparency* (specified in Paragraph 2). The need for *environmental integrity* to guarantee effectiveness in the actions undertaken under A6PA appears of utmost importance, especially in the context of mitigation. This is confirmed by the prominence given to avoidance of double counting in Paragraph 2 and further reiterated in the description of rules governing the mechanism (Paragraph 5). The call for *transparency*, introduced in Paragraph 2 and pervading the whole text, represents a clear imperative for any action implemented under the A6PA framework. Independence in relation to transparency is stressed in Article 13 of the PA.

Discussing the unique relationship that may eventually manifest between the governance regimes promoted through A6PA and the general framework of the PA and the UNFCCC represents the final step to fully understand the architecture and functioning of the model.[8] *Public and private entities* represent a key term in this discussion. Appearing twice in the text (in Paragraphs 4 and 8), the term holds a high significance for A6PA and its interrelated experimental governance options. The term is used to stress the objective of increasing participation of entities other than parties as part of the overall implementation of A6PA. While it appears evident that non-party entities are meant to become primary stakeholders for the development of approaches to mitigation outcomes' exchange and measures introduced under the framework of A6PA, it is worth adding some further considerations on their role in the general implementation of governance regimes instigated through A6PA.

The text does not clarify definitively whether 'some parties voluntarily collaborating' under the A6PA framework correspond only to parties as generally defined in international treaties such as the PA. On the one hand, the relevance given to the NDCs within the text (mentioned above) may suggest that the parties active in the regime of A6PA should in principle be part of the group of parties active in the PA and traditionally interpreted as countries or jurisdictional territories (e.g. the European Union). On the other hand, the urgency of increased participation of *public and private entities*, supported by the freedom given to parties in *authorising* (term used twice in the text, in Paragraphs 3 and 4, with the latter directly referring to the *authorisation* given by parties for public and private entities' participation)

[7] See Chaps. 6 and 7 of this book.

[8] For a detailed analysis of this relationship see Sect. 8.2 of Chap. 8 of this book.

actions within the structure of A6PA, leaves a window of opportunities for increasing engagement of entities (stakeholders) other than traditional parties which may indeed become prominent members of the cooperative systems established eventually (see Paragraph 2).

This interpretation has been recently supported by the decision of the Canadian Government to give its provinces freedom and flexibility in determining their models to meeting the emissions objectives stated by the country's NDC (Canadian Government 2016). As a consequence of this decision, Canadian provinces, which can be classified as public authorities other than parties, may eventually become members of cooperative groups constituted under A6PA. Similar decisions taken by other governments concerning both *public and private entities* may ultimately lead to the establishment of a wide variety of cooperative groups under A6PA, composed of any category of stakeholders.

To complete the discussion on the relationship between the governance regimes suggested by A6PA and the general system of the UNFCCC, an understanding of the links between the structure of A6PA and the Conference of Parties serving as the meeting of the Parties to the Paris Agreement (CMA) is required. On one hand, the text appears to align with the general norms governing the implementation of the PA. For instance, it will be for the CMA to adopt the *rules, modalities and procedures* (another significant term appearing once in Paragraph 4) required by A6PA and its *mechanism*, thus confirming that this adoption of norms will not be delegated to those parties that will decide to cooperate under the framework of A6PA. On the other hand, the role of the CMA in the overall implementation of A6PA is mentioned in several parts of the text. For instance, Paragraph 4 recalls that the *mechanism* will be "established under the authority and guidance of the CMA for use by parties on a voluntary basis", thus providing an amount of autonomy in its establishment (implemented *under the authority and guidance* of but not directly by the CMA). The Paragraph also states that supervision of the mechanism will be the responsibility of a "body designated by the CMA", without specifying its nature and relationship to the CMA. Based on these legal premises, the *body* may eventually act externally and independently of the UNFCCC, therefore offering a multitude of interpretations, which will be further analysed in Chap. 8 of this book. Finally, the use of the term *consistent with* instead of the more legally binding *in accordance with* (adopted 13 times in other parts of the PA but never within A6PA) in Paragraph 2, which aims to define the relationship between the actions resulting from the A6PA governance regimes and the CMA, further justifies the opinion that the structure and functioning of the regimes eventually developed through A6PA may act parallel to and with a high degree of autonomy from the UNFCCC system.[9]

A detailed discussion of the term *shall* concludes the analysis of A6PA terminology. Appearing nine times, the term represents a fundamental element of the A6PA's design. The use of the term *shall* in international legal treaties, including the PA, imparts a binding meaning to whatever it refers to. The term is used 117 times in the 29 Articles of the PA and aims to stress its binding significance. This also holds true

[9] See also Chap. 8 of this book.

for its recurring presence in A6PA. *Shall* specifies that, despite the voluntary nature of any cooperation resulting from the application of the framework of A6PA, such cooperation requires alignment to a specific set of binding rules and aims. By looking at the use of *shall* within A6PA, it is possible to identify these rules and aims as follows:

- When engaging in their cooperative activities, involved parties *shall*: "promote sustainable development and ensure environmental integrity and transparency" (Paragraph 2).
- And *shall*: "apply robust accounting to ensure, inter alia, the avoidance of double counting" (Paragraph 2).
- The use of ITMOs *shall*: "be voluntary and authorized by participating parties" (Paragraph 3).
- The *mechanism* of A6PA *shall*: "be supervised by a body designated by the CMA" (Paragraph 4).
- And it *shall* aim: "(a) to promote the mitigation of greenhouse gas emissions while fostering sustainable development; (b) to incentivize and facilitate participation in the mitigation of greenhouse gas emissions by public and private stakeholders authorized by a party; (c) to contribute to the reduction of emission levels in the host party, which will benefit from mitigation activities resulting in emission reductions that can also be used by another Party to fulfil its nationally determined contribution; and (d) to deliver an overall mitigation in global emissions" (Paragraph 4).
- Mitigation resulting from applying the *mechanism shall*: "not be used to demonstrate achievement of the host Party's NDC if used by another Party to demonstrate achievement of its NDC" (Paragraph 5).
- The CMA *shall*: "ensure that a share of the proceeds from activities under the *mechanism*… is used to cover administrative expenses as well as to assist developing country Parties that are particularly vulnerable to the adverse effects of climate change to meet the costs of adaptation" (Paragraph 6).
- The CMA *shall*: "adopt rules, modalities and procedures for the *mechanism*" (Paragraph 7).
- Non-market approaches supported by the Article *shall* aim to: "(a) promote mitigation and adaptation ambition; (b) enhance public and private sector participation in the implementation of NDCs; and (c) enable opportunities for coordination across instruments and relevant institutional arrangements" (Paragraph 8).

While its absence from Paragraph 1 serves to reaffirm the voluntary nature of actions taken under the umbrella of A6PA, the widespread use of *shall* within the text provides a significant guideline for whoever aims to formalise new climate governance regimes based upon the provisions of A6PA. The climate governance regime proposed in this book (the Mitigation Alliance—MA, introduced in Chap. 3) refers to this guiding aspect as a key tool for its definition and design.

4.4 A Transformational Club Under Article 6 of the Paris Agreement: The Mitigation Alliance

Resulting from the features introduced in Sects. 4.2 and 4.3 above as well as the legal framework of the PA, any climate governance regime framed under A6PA would be required to meet some specific obligations while also enjoying a wide variety of opportunities. This combined mix of duties and rights can effectively trigger unique experimental governance regimes in current and future climate policy.

These features can be synthesized into ten provisions for the basis of any governance regime implemented through A6PA:

1. Parties, as well as non-state entities authorised by their reference parties, can engage in minilateral voluntary cooperation, aimed at higher ambitions in climate change response.
2. By cooperating, parties and other stakeholders can design *ad hoc* rules to determine their cooperation structure and functioning in accordance with A6PA principles, aims and guidelines.
3. To meet A6PA criteria and PA objectives, cooperative members shall *a priori* determine a common Net Zero Carbon (NZC) emissions target and shall agree on its distribution in form of assigned share of mitigation burden, or quota.[10]
4. In its implementation, the cooperation shall guarantee transparency and environmental integrity, thus avoiding, *inter alia*, double counting of mitigation outcomes.
5. The cooperation will adopt a single mechanism to certify mitigation outcomes required to meet its target.
6. The mechanism will be applied to any approach for exchanging mitigation outcomes adopted within the cooperation system including, *inter alia*, market and non-market approaches.
7. Overseen by a designated Body, application of the mechanism shall contribute to effectiveness and equity of mitigating actions, while supporting adaptation and resilience strategies.
8. Actions and strategies implemented through the cooperation shall guarantee sustainable development in all its aspects including environmental, social and economic areas.
9. In principle, inclusion in the cooperation is open and voluntary, but its founding partners may formulate norms to regulate the inclusion process.
10. The founding partners can design exclusion norms both concerning partners failing to meet their quotas and entities external to the cooperative regime.

Bearing in mind these 10 founding provisions, it is possible to establish several links with the game theory, the experimentalist governance model and coalition formation literature (Sabel and Zeitlin 2011; Caparrós 2016). International

[10] Quota is used in this Chapter and more in general in this book as synonym of 'assigned share of mitigation burden'.

Environmental Agreements (IEA) studies based on cooperative game theory (Chander and Tulkens 1997), which represent the origins of the climate clubs literature discussed in Chap. 3 of this book (Stewart et al. 2013), may provide the ideal framework for implementing experimental climate governance regimes as envisaged in A6PA.

This book, whose ultimate objective is to propose an experimental governance regime corresponding to the criteria defined by A6PA, identifies in these links between A6PA on the one hand, and cooperative game theory and climate clubs literature on the other, the founding elements for the implementation of its governance proposal. Therefore, before sketching the skeletal elements of the architecture of the proposed governance regime, this Chapter proposes a synthesis aimed at pinpointing the key elements of contact between the identified provisions of A6PA and cooperative game theory and climate clubs literature.

While describing the major conditions for a successful climate club, Nordhaus (2015) defines the club approach as a 'top-down approach' comparing it to Bretton Woods as an ideal model (Nordhaus 2015, p. 1344). Compared to the 'pledge and review' approach envisaged by the PA, any structure eventually resulting from A6PA appears top-down. Moreover, all four conditions identified by Nordhaus (2015, p. 1340) can be directly or indirectly identified within the above-listed ten provisions of A6PA. These links between Nordhaus' perspectives and A6PA are confirmed when the analysis is extended to include other scholars.

First, the concept of 'voluntary cooperation', which is the principle of A6PA, represents one of the keystones of climate clubs literature (Hale 2011; Das 2015; Nordhaus 2015; Orsato et al. 2015; Potoski 2015). Weischer et al. (2012) introduce the concept of 'bringing in', which suggests "that groups of countries who are willing to undertake particular commitments or actions agree to have those negotiated, recognized and monitored within the UNFCCC" (Weischer et al. 2012, p. 191), which closely resembles the original significance of A6PA. Secondly, A6PA, and the PA in general, can satisfy the need for clubs' international legitimacy, reiterated by many of the most prominent scholars in clubs literature (Eckersley 2012; Das 2015; Falkner 2016). This element is further emphasised by Weischer et al. (2012), who explicitly stress the need for coordination under an international framework for transformational climate clubs, and by Stewart et al. (2013), while stating that linking clubs' regimes to the UNFCCC process "in a way that respects the need for multilateralism in addressing the global character and consequences of climate change may help defusing criticism" of the regimes themselves (Stewart et al. 2013, p. 10).

Higher ambition represents another element shared between the clubs literature and the architecture of A6PA. Authors such as Das (2015) and Falkner (2016) identify in the clubs' need for higher ambition in mitigation an opening for legitimising great power cooperation in the context of the existing international regime (the PA), which confirms the meaning and significance of A6PA. In addition, Keohane et al. (2015) identify in the clubs the resource for broadening participation and ambition in mitigation, while Weischer et al. (2012) include ambition in the list of essential features of transformational clubs. Unsurprisingly, this list includes four elements that correspond with A6PA: speed, ambition, participation and equity. As a consequence of higher ambition and based upon the pioneering work of Carraro and Siniscalco (1993), some clubs scholars stress the need for clubs to determine quantified aggregate and individual mitigation targets for their members (Weischer et al. 2012; Stewart et al. 2013; Victor 2015), which provides a further link to the analysis of A6PA in the previous sections above.[11] Aside from the need for higher ambition, clubs scholars in the past also emphasised the need for clubs to be designed with rules and features that fit the social, economic, and policy circumstances in which they operate (Potoski and Prakash 2009), again linking clubs literature with the significance of NDCs in A6PA.

Climate clubs literature also discusses the use of certification *mechanisms* similar to the one introduced in A6PA as enhancing tools for action within the clubs. More specifically, some scholars favour certification to facilitate carbon market approaches (Keohane et al. 2015), while others emphasise their possible effect of enhancing transparency (Reid and Toffel 2009; Potoski 2015). While analysing the functioning of the *mechanism*, Keohane et al. (2015) also emphasise the significance of avoiding double counting, aligning once more with the principles of A6PA.[12]

If the considerations listed above confirm the links between climate clubs literature and A6PA, experience and effort in developing the former can contribute substantially towards solving the yet open issues of the latter. Widely discussed in climate clubs literature is only the need but also the effective implementation of measures for transparency, accounting and monitoring, verification and reporting (MRV), which represent core elements of both A6PA and the whole PA (Potoski and Aseem 2005; Stewart et al. 2013; Keohane et al. 2015; Falkner 2016).[13] Climate clubs scholars also analysed in depth the criteria for membership and non-compliance and discuss the legal implications of similar criteria in clubs' regimes (Eckersley 2012; Weischer et al. 2012; Stewart et al. 2013; Bulkeley et al. 2014; Keohane et al. 2015; Falkner 2016).[14]

Finally, the issue of excludable benefits assigned to the clubs' members, which represents by far one of the most extensively discussed elements of the clubs literature (Weischer et al. 2012; Stewart et al. 2013; Keohane et al. 2015; Nordhaus 2015; Orsato et al. 2015; Falkner 2016), may well help define norms aimed at attracting stakeholders towards the experimental climate governance models initiated through A6PA.[15] These discussions may also help define approaches, possibly based upon different categories of benefits, which can be implemented within an A6PA governance model.[16] In their analysis of risks of clashing with international jurisdictions,

[11] See also Chap. 5 of this book.

[12] See also Chap. 6 and Sect. 8.10 of Chap. 8 of this book.

[13] Extended comments on the significance of MRV can be found in Chaps. 5, 6, 7 and 8 of this book.

[14] See also Chap. 8 of this book.

[15] See also Chaps. 9 and 10 of this book.

[16] See also Chap. 7 of this book.

the studies on excludable benefits may also help understand and prevent possible contradictions between the new regimes instigated by A6PA and other current legal frameworks, such as the World Trade Organization (WTO).[17]

4.5 Conclusions: Drafting the Architecture of the Mitigation Alliance

As apparent from the discussion in the previous sections of this Chapter, the experimental governance model proposed in this book builds on the findings gleaned from an analysis of A6PA and analogies drawn from climate clubs literature. Consequently, the proposed model, referred to hereafter as the Mitigation Alliance (MA), comprises ten components, which can be summarised as follows:

1. In accordance with Paragraph 1 of A6PA and with voluntary participation as envisaged in the clubs literature, the MA will be composed exclusively of members who voluntarily decide to cooperate and of those entities authorised by the members.[18]
2. Following Paragraph 7 of A6PA and the general understanding of clubs literature, the members' composition will vary according to the rules, modalities and procedures that will define its structure set up by the MA.[19]
3. In accordance with Paragraphs 1, 4 and 8 of A6PA and in line with the objectives of clubs literature, the MA will define a quantified, aggregate mitigation target that will be aligned with the objectives of the PA.[20]
4. Again in accordance with Paragraphs 1, 4 and 8 of A6PA and in line with the objectives of clubs literature, the MA will design a system for assigning shares of the quantified target to its members, based upon burden sharing criteria.[21]
5. In response to the analysis of Paragraphs 4, 5, 6 and 7 of A6PA according to clubs literature perspectives, the MA will adopt a single mechanism to certify the mitigation outcomes required to meet the agreed target.[22]
6. In accordance with Paragraphs 2, 4 and 8 of A6PA, Article 5 and findings of clubs literature, the MA will establish a variety of approaches to be adopted by its members to exchange mitigation outcomes for their assigned shares of mitigation, hence meeting the MA aggregate target.[23]
7. In order to satisfy the criteria included in Paragraphs 2, 4 and 5 of A6PA and following the suggestions of clubs literature, the MA will include tools to

[17] See also Sect. 8.3 of Chap 8 of this book.
[18] See also Chap. 8 of this book.
[19] See also Chap. 8 of this book.
[20] See also Chap. 5 of this book.
[21] See also Chap. 5 of this book.
[22] See also Chap. 6 of this book.
[23] See also Chap. 7 of this book.

guarantee effectiveness, equity, environmental integrity and transparency in every aspect of its implementation.[24]

8. In accordance with Paragraphs 2, 6 and 8 of A6PA and based upon the benefits criteria of clubs literature, the MA will provide direct support for adaptation and resilience actions undertaken by its members.[25]
9. Based upon Paragraphs 1, 2, 3, 4, 5 and 8, supported by the decisions related to the implementation of Paragraph 7 of A6PA and inspired by clubs literature, the MA governance will be characterised by a multi-level, hybrid model aligned with the UNFCCC and PA principles.[26]
10. As a consequence of its implementation, the MA will offer a wide variety of excludable benefits for its members, which will extend far beyond the non-excludable benefit of mitigating global emissions.[27]

These ten key elements form the basis of the in-depth analysis of the proposed MA governance model that follows in the next Chapters of this book.

References

Bodansky, D. (2016). The legal character of the Paris agreement. *RECIEL, 25*(2), 142–150.

Bulkeley, H., Andonova, L., Betsill, M. M., Compagnon, D., Hale, T., Hoffmann, M. J., Newell, P., Paterson, M., Roger, C., & VanDeveer, D. D. (2014). *Transnational climate change governance*. New York: Cambridge University Press.

Canadian Government. (2016). Government of Canada announces pan-Canadian pricing on carbon pollution. Canadian Government press release, 3 October 2016. http://news.gc.ca/web/article-en.do?nid=1132149

Caparrós, A. (2016). The Paris agreement as a step backward to gain momentum: Lessons from and for theory. *Revue d'Economie Politique, 126*(3), 347–356.

Carraro, C., & Siniscalco, D. (1993). Strategies for the international protection of the environment. *Journal of Public Economics, 52*(3), 309–328.

Chander, P., & Tulkens, H. (1997). The Core of an economy with multilateral environmental externalities. *International Journal of Game Theory, 26*(3), 379–401.

Clémençon, R. (2016). The two sides of the Paris climate agreement: Dismal failure or historic breakthrough? *Journal of Environment and Development, 25*(1), 3–24.

Das, K. (2015). Climate clubs: Carrots, sticks and more. *Economic & Political Weekly, 50*(34), 24–27.

Eckersley, R. (2012). Moving forward in the climate negotiations: Multilateralism or Minilateralism? *Global Environmental Politics, 12*(2), 24–42.

Falkner, R. (2016). A minilateral solution for global climate change? On bargaining efficiency, club benefits and international legitimacy. *Perspectives on Politics, 14*(1), 87–101.

Hale, T. (2011). A climate coalition of the willing. *Washington Quarterly, 34*(1), 89–101.

[24] See also Chaps. 5, 6, 7 and 8 of this book.

[25] See also Chaps. 8 and 10 of this book.

[26] See also Chap. 8 of this book.

[27] See also Chaps. 9 and 10 of this book.

IPCC. (2014). Climate Change 2014: Synthesis Report. Contribution of Working Groups I, II and III to the Fifth Assessment Report of the Intergovernmental Panel on Climate Change [Core Writing Team, R. K. Pachauri and L. A. Meyer (Eds.)]. IPCC, Geneva, 151 pp.

Keohane, N., Petsonk, A., & Hanafi, A. (2015). Toward a club of carbon markets. *Climatic Change*. doi:10.1007/s10584-015-1506-z.

Marcu, A. (2016, January). *Carbon market provisions in the Paris agreement (Article 6)* (CEPS Special Report 128).

Nordhaus, W. (2015). Climate clubs: Overcoming free-riding in international climate policy. *American Economic Review, 105*(4), 1339–1370.

Orsato, R. J., de Campos, J. G. F., Barakat, S. R., Nicolletti, M., & Monzoni, M. (2015). Why join a carbon club? A study of the banks participating in the Brazilian "Business for Climate Platform". *Journal of Cleaner Production, 96*, 387–396.

Potoski, M. (2015). Green clubs in building block climate change regimes. *Climatic Change*. doi:10.1007/s10584-015-1517-9.

Potoski, M., & Aseem, P. (2005). Covenants with weak swords: ISO 14001 and firms' environmental performance. *Journal of Policy Analysis and Management, 24*(4), 745–769.

Potoski, M., & Prakash, A. (2009). *Voluntary programs: A club theory approach*. Cambridge, MA: MIT Press.

Rajamani, L. (2016). Ambition and differentiation in the 2015 Paris agreement: Interpretative possibilities and underlying politics. *International and Comparative Law Quarterly, 65*, 493–514.

Reid, E. M., & Toffel, M. W. (2009). Responding to public and private politics: Corporate disclosure of climate change strategies. *Strategic Management Journal, 30*(11), 1157–1178.

Sabel, C. F., & Zeitlin, J. (2011). Experimentalist governance. In D. Levi-Faur (Ed.), *The Oxford handbook of governance*. Oxford: Oxford University Press.

Stewart, R., Oppenheimer, M., & Rudyk, B. (2013). A new strategy for global climate protection. *Climatic Change, 120*, 1–12.

Van Asselt, H. (2016). International climate change law in a bottom-up world. *Questions of International Law, 26*, 5–15.

Van den Bergh, J. C. J. M. (2016). Rebound policy in the Paris agreement: Instrument comparison and climate-club revenue offsets. *Climate Policy*. doi:10.1080/14693062.2016.1169499.

Victor, D. G. (2015, January). The case for climate clubs. *E15 Initiative Thinking Piece*.

Waisman, H., Torres Gunfaus, M., Spencer, T., & Marquard, A. (2016). Emerging from Paris: Post-2015 process, action and research agenda. IDDRI working paper, Paris.

Weischer, L., Morgan, J., & Patel, M. (2012). Climate clubs: Can small groups of countries make a big difference in addressing climate change? *RECIEL, 21*(3), 177–192.

Part II
Architecture of the Mitigation Alliance

Chapter 5
The Mitigation Alliance Target and Its Distribution

Michele Stua and Michele Coulon

Abstract Misalignment between the Paris Agreement (PA) objectives and its implementation tools represents the biggest challenge for its implementation. The compromise between top-down urgencies, such as the 2/1.5 °C target, and bottom-up stances, such as the Nationally Determined Contributions (NDCs) pledge-and-review system removed years of efforts for identifying effective and equitable systems to combat climate change, threatening the same signification of the Common But Differentiated Responsibilities and Respective Capabilities (CBDR-RC) principle. The Mitigation Alliance (MA) proposed in this book offers a feasible alternative to the current system. It applies to entities that voluntarily collaborate under the framework of the PA's Article 6 (A6PA) and entails the establishment of an aggregate Net Zero Carbon (NZC) target for the MA and a technical, flexible distribution of the corresponding mitigation burden. While the NZC is based on scientific evidence, the hereby-proposed distribution system represents a key innovation. By introducing a dynamic formula for distributing the mitigation burden (or quota) among its members, the MA aligns ambition with the PA's objectives while taking full account of the CBDR-RC principle.

Keywords Paris Agreement (PA) objectives • Common but Differentiated Responsibilities and Respective Capabilities (CBDR-RC) • Nationally Determined Contributions (NDCs) • Article 6 of the PA (A6PA) • Net Zero Carbon (NZC) • Mitigation burden

5.1 Introduction

Setting global limits on carbon[1] emissions in the form of target limits has been the priority in climate change policy since its very early stages, together with the redistribution of these limits across parties. Both are complementary elements of a single, rational perspective (Babiker and Eckaus 2002).

[1] Carbon Dioxide (CO_2) is the most common Greenhouse Gas (GHG) and is often used as a reference unit to express levels of GHG emissions (expressed in CO_2 equivalent, or CO_2eq). In the context of this Chapter and, more in general, of this book the term carbon, expressing CO_2, and the term GHG are synonyms, hence carbon is used as single term to express GHG.

M. Stua, *From the Paris Agreement to a Low-Carbon Bretton Woods*,
DOI 10.1007/978-3-319-54699-5_5

Both climate science and policy-making have concentrated on finding a solution acceptable to all members of the international community that would nevertheless take account of the global climate-related needs. The outcomes of these efforts are reflected in the Paris Agreement (PA), where the target and its distribution are addressed with two completely different governance models, but with an unsatisfactory result.

By representing the ultimate compromise on climate change for virtually the entire world, the PA contains elements designed by a multitude of actors aimed at influencing an even larger set of social, economic and political domains. This variety is reflected also in the different governance models embedded in the articles of the Agreement. The text of the PA thus appears as a mix of top-down and bottom-up models, which often seem to contradict each other. The definition of the global target and its redistribution stands as a perfect example of such a contradiction.

On the one hand, the PA adopted a top-down model in setting ambitious overall objectives, expressed as ceilings for global temperature increase compared to pre-industrial levels, thus ultimately resolving the carbon target issue. Inspired by the scientific evidence of the 5th report of the Intergovernmental Panel on Climate Change (IPCC) (IPCC 2014), a solid architecture was designed for the PA to specify its ultimate objectives and to transform them into binding aggregate targets for all its participants. This ultimately resolved the issue of how to define the target.

On the other hand, the PA adopted a bottom-up model based on the pledge-and-review system in order to redistribute and implement the objectives. By adopting a pledge-and-review system, whereby decisions concerning a quota of responsibility are passed on to those responsible for meeting the quota, it was intended to stimulate greater participation by the international community members. Yet this approach abandons binding mitigation systems (Clémençon 2016) by not offering any direct link between the target and its redistribution among the parties and therefore fails to stimulate greater participation.

This failure is evident in the analysis of pledges submitted to the United Framework Convention on Climate Change (UNFCCC) by its parties between the second half of 2015 and the first half of 2016. Proposed in the form of (Intended) Nationally Determined Contributions ((I)NDCs), the pledges should have, on aggregate, approximated a mitigation pathway in line with the mitigation target as defined by the overall objectives of the PA. Preliminary studies on the (I)NDCs show that, despite some positive changes compared to business-as-usual scenarios (den Elzen et al. 2016), the sum of the parties' mitigation ambitions introduced in their pledges falls well below those required by any of the mitigation pathways identified to date (Peters et al. 2015; du Pont et al. 2016; Rogelj et al. 2016; van Asselt 2016; van den Bergh 2016). In addition, Peters et al. (2015) and du Pont et al. (2016) stress how the (I)NDCs in their current form also fail to meet the minimum equity criteria which, through the Common But Differentiated Responsibilities and Respective Capabilities (CBDR-RC) principle, are fundamental to the PA.

Guaranteeing an effective and equitable distribution of responsibilities among the parties is therefore necessary in order to overcome the inherent shortcomings of the pledge-and-review approach. The Nationally Determined Contributions (NDCs) defined by the PA differ from the (I)NDCs of the draft version in that the omission of the word 'Intended' conveys the explicitly binding meaning of the NDCs. Nevertheless, the link between cycles of NDCs and the 2/1.5 °C target is only implicit in the PA (Waisman et al. 2016), therefore leaving unsolved issues.

Yet, despite the lack of explicit solutions within the PA, some elements contained in it can facilitate the development of a process aimed at addressing this lack. First, the global stocktake described in Article 14 of the PA may indeed facilitate the enhancement of parties' NDCs. The global stocktake represents an instrument adopted by the Conference of the Parties serving as the meeting of the Parties to the Paris Agreement (CMA) to take stock of the PA implementation at regular intervals (every 5 years starting from 2023). Its findings are intended to provide a benchmark to help parties update and enhance their actions in accordance with the PA provisions, as well as to intensify international cooperation for climate action. Although this function is not explicitly defined in its implementation rules, global stocktake shall therefore become a reference for stimulating the alignment of parties' NDCs to the overall PA objectives (Waisman et al. 2016). Furthermore, some authors identify in the inclusion of equity within the text of Article 14 an open door for a dialogue on equitable burden sharing (Rajamani 2016).

Secondly, and most importantly, Article 6 of the PA (A6PA) presents a unique framework for experimenting with new governance models.[2] It provides opportunities to 'bridge the gap' between the PA objectives and its current pledge-and-review implementation system. The governance model suggested in this book, the Mitigation Alliance (MA), is a mix of top-down and bottom-up perspectives consistent with the overall structure of the PA. But before the potential of the MA to solve the current inconsistency between the overall objectives of the PA and its implementation tools can be established, it is first necessary to discuss several concepts related to the target itself and its distribution. This chapter analyses each of these concepts in depth, with references to recent findings from climate policy literature for support.

A MA limited to entities that voluntarily decide to cooperate, following the rules laid down by A6PA, will find itself free to determine an aggregate target, which can be understood as the MA's common responsibility. To be consistent with the PA mandate, this responsibility shall correspond to the MA's aggregate share of the overall PA objectives.

To guarantee this consistency, the MA target shall be determined as part of a global mitigation pathway in line with the scenarios based on the IPCC findings (IPCC 2014). This implies adopting recommendations formulated by the most recent scientific studies on temperature stabilisation scenarios and combining them with corresponding pathways related to the IPCC findings, as discussed in Sect. 5.2 of this Chapter.

Representing the aggregate share of responsibilities of the MA members in relationship to the identified pathway, the target should be submitted to the UNFCCC and its associated organizations in the form of a Plurilaterally Determined

[2] See Chap. 4 of this book.

Contribution (PDC). Although the PA does not explicitly envisage the design of PDCs, it does not include obstacles to their establishment and implementation either (Waisman et al. 2016). As part of a mitigation pathway, the PDC may contain both a long-term target and its distribution timetable. If organized in the form of 5-year timetables, the PDC may be consistent with the rules governing the pledge-and-review NDC system hence satisfying the mandate of A6PA towards them.[3]

Apart from an aggregate target, the MA also requires the adoption of an allocation-based, 'burden sharing' (Böhringer and Welsch 2006) to redistribute its target in an effective and equitable way. This is necessary to help overcome limitations associated with the pledge-and-review system of the PA. The burden sharing literature has developed extensively over the years (Sijm et al. 2001; Böhringer and Welsch 2006; Luderer et al. 2012; Bretschger 2016), leading to a variety of redistribution proposals typically categorised as top-down. Such approaches may thus appear in contrast to the PA architecture. Nevertheless, Paragraph 1 of A6PA offers the freedom and opportunity to develop experimental governance models, such as the MA, capable of stimulating innovative mitigation regimes and features, in this including the redistribution of responsibilities or 'burden sharing'. Section 5.3 of this Chapter discusses the 'burden sharing' system proposed within the MA.

In summarising the findings of this Chapter, conclusive Sect. 5.4 points out how the established MA mitigation target represents the foundation for the demand of mitigation outcomes within the same MA.

5.2 Adoption of an Aggregate Target in Line with the Paris Agreement Objectives

Adopting an aggregate, or common, binding carbon target within the MA is necessary not only to engage the MA in the process of meeting the PA objectives; it also implies a series of benefits that may encourage a variety of entities to join the MA. Carbon mitigation represents just the most direct (non-excludable) benefit of committing to the target. While the nature of this benefit is analysed in detail in other sections of this book, this section demonstrates how, by setting a target based on a defined, minimum amount of mitigation outcomes to be reached within a specified timetable, it is possible to guarantee additional (excludable) benefits to MA members.[4]

To determine the aggregate target of the MA, which ultimately corresponds to the share of the PA objectives under the responsibility of the MA, it is first necessary for these objectives to be quantified in carbon emission mitigation (or simply mitigation) terms. In principle it is also necessary to understand how an objective expressed as a temperature ceiling can be translated into a quantified amount of

[3] The interrelationship between the PDC and the NDC system is explored further in Chap. 8 of this book.

[4] See below Chaps. 9 and 10 of this book.

mitigation for a determined period, or timetable, using the measure of carbon. A significant amount of recent scientific literature facilitates this translation.

Drawing on the findings of the 5th report of the IPCC (IPCC 2014), a broad debate on how to reach and stabilise global average temperatures at a ceiling no higher than 2 °C compared to the pre-industrial level has arisen in recent years. By including the ceiling (Article 2, Paragraph 1) and its stabilisation over time (Article 4, Paragraph 1) among its objectives, the PA explicitly binds the signing parties to adopt pathways in line with the IPCC findings. As a consequence, the scientific IPCC report findings can now offer effective solutions for policy-makers seeking to meet their binding objectives.

An ongoing debate on the rationale of and pathways for the 2 °C stabilisation, where various, non-excludable proposals and solutions are analysed, requires the involvement of both scholars and policy-makers. The aim of this Chapter is not to review the whole range of proposals currently on the table; rather, the Chapter focuses more on specific governance models and the literature that relates most closely to key elements of these models.

The scientific discussion concerning the 5th IPCC report and its translation into policy action has been generally defined as 'carbon' literature. The carbon budget represents a target quantity of cumulative carbon emissions in line with the findings of the IPCC and its pathways to achieve the 2 °C temperature objective. The cumulative carbon budget as proposed by the IPCC identifies an overall amount of CO_2 that, once released into the atmosphere by anthropogenic actions, would effectively lead to a maximum 2 °C global average temperature increase compared to the pre-industrial level.

Representing the concentration of CO_2 in the atmosphere, expressed in parts per million by volume (ppmv), the carbon budget has both advantages and disadvantages as a measurement tool to determine required policy action. Usually calculated in relationship to a baseline period between 1861 and 1880 (Millar et al. 2016), the carbon budget is typically based on a concentration target of 450 ppmv (IPCC 2014). The concentration target is then translated into an overall target for cumulative emissions expressed in gigatonnes of CO_2 (CO_2 Gt). By subtracting the already 'spent' emissions from the original cumulative amount, it is possible to quantify the volume of CO_2 emissions still 'available' before breaching the 450 ppmv target.

The use of a cumulative carbon budget for climate policy promises a series of advantages, ranging from its immediateness to its potential as reference for the distribution of emissions allowances (Ohndorf et al. 2015). As a consequence, a variety of studies have focused on the definition of the overall cumulative budget and the share that can still be 'consumed'. Challenged by uncertainties over the physical climate response and the role of non-CO_2 Greenhouse Gases (GHG), the studies often differ in their final recommendations and generally offer a significant range of results based on probabilities (expressed in percentages), rather than defining a specific amount of 'allowable' emissions (Ohndorf et al. 2015; Tavoni et al. 2015; Millar et al., 2016; Rogelj et al. 2016; Zaman et al. 2016).

Uncertainties relating to the adoption of a cumulative carbon budget limit its effectiveness as a direct tool with which to determine a global mitigation target, therefore compromising its efficacy in defining the corresponding target for the MA. Nevertheless, studies linked to the carbon budget have recently evolved towards new approaches for the definition of the mitigation target. More specifically, the carbon budget approach has been used to define specific budgets (based on the so-called peak years) for different countries or regions (Wei et al. 2014), thus supporting an equitable distribution of responsibilities, as well as to propose the Net Zero Carbon (NZC) philosophy (Clémençon 2016; Millar et al. 2016). The NZC philosophy assumes that "no matter which approach is taken, the CO_2 budget for keeping warming to below 2 °C always implies stringent emission reductions over the coming decades and net zero CO_2 emissions in the medium to long term" (Rogelj et al. 2016, p. 251). NZC helps overcome uncertainties linked to the application of different models for calculating the carbon budget.

To meet the overall objectives of the PA, Clémençon (2016) specifies 44 Gt of CO_2 equivalent as the quantity of mitigation required to reduce the annual level of global emissions expected by 2020 (56 Gt) to a 2050 limit in line with the NZC pathway (8 Gt). Clémençon (2016) also identifies 2090 as the ultimate deadline to meet a definitive NZC emission level. By adopting the NZC philosophy, it is possible to define both the intermediate (2050) and the final mitigation targets of the MA. The latter will, of course, correspond to the entirety of the current, aggregate MA carbon emissions, while the former can be given as follows:

$$MAt = \frac{44}{56} MAb$$

where *MAt* represents the 2050 intermediate mitigation target for the MA expressed in quantities of mitigation outcomes, and *MAb* represents its carbon emissions level in the baseline year (identified by Clémençon as 2020). As the aggregate level of emissions within the MA is supposed to decline from the baseline year onwards, *MAb* also represents the aggregate peak year for MA.

Based on the NZC findings, aligning the MA target to the objectives of the PA would therefore require a consistent amount of mitigation, defined in both medium- and long-term targets. If agreed by the MA members, the amount would therefore represent an overall demand for mitigation outcomes. Introducing a consistent, defined amount of demand for mitigation also calls for an equal, corresponding supply.

The demand/supply dynamics[5] established by the target setting represent the foundation for additional benefits capable of attracting new members and keeping existing members within the MA. Yet, before discussing such benefits in detail,[6] other aspects concerning target management must be clarified, such as the temporal distribution of the identified target. While organizing the target into 5-year timetables appears the most obvious solution to keep the MA regime in line with the

[5] See Also Chaps. 6 and 7 of this book.

[6] See Chaps. 9 and 10 of this book.

similarly organized approach chosen for the PA, specifying the exact shares of the MA target to be achieved by the defined timetables is not self-evident.

The shares of the MA target may be equally distributed over the defined timetables, or may vary in magnitude across them. MA members shall have discretion in determining the exact timetable for the distribution of such shares, based on various rationales. Nevertheless, it would be useful for such distribution to be defined *a priori*, in order to offer a steady degree of transparency for the mitigation pathway identified by the MA. Additionally it is relevant to remember that, by corresponding to the MA's PDC the MA target requires to align to the mandates of in order to be Article 4 of the PA concerning NDCs. As a consequence, any share of target defined for a timetable shall be equal or higher than the previous share.[7]

By adopting a consistent and determined target, and by defining its distribution in shares over fixed timetables, the MA would then sow the seeds for the establishment of the new governance model proposed in this book. For the model to be completely effective, the MA would be required to take a further step by establishing a burden sharing system among its members. To be consistent with the PA, the system should be aligned to the CBDR-RC principle.

5.3 CBDR-RC Burden Sharing Within the Mitigation Alliance

Aimed at ensuring effectiveness and equity in designing the distribution of mitigation obligations (Deleuil 2012), CBDR-RC has acted as a key principle for the determination of burden sharing systems since its inception (UNFCCC 1992, Article 3, Paragraph 2).[8] The PA confirmed the CBDR-RC significance by including it as one of its guiding principles (Article 2, Paragraph 2).[9]

The main objective of the CBDR-RC concept is to address climate change efficiently, while taking into account the different capabilities and differing responsibilities of individual countries. Climate policy literature emphasises that the CBDR-RC principle requires the simultaneous satisfaction of its equity and effectiveness components in order to meet its objective (Markandya 2011; Garibaldi 2014). Hence a burden sharing system in line with the PA is needed in order to satisfy the two requirements embedded within the CBDR-RC simultaneously.

Despite its relevance, reiterated by the PA, a fully explanatory definition of the CBDR-RC principle has not yet been found. Equity represents by far the most

[7] Paragraph 3 of Article 4 of the PA.

[8] See also Chap. 2 of this book.

[9] Yet the strict original interpretation of CBDR-RC as adopted in the PA's predecessor, the Kyoto Protocol, where CDBR-RC served at divide parties responsible for mitigation from exempted ones required reinterpretation in a more nuanced fashion to help operationalize equity and help guide the fashioning of a commitment in a regime 'applicable to all' in the PA (Winkler and Rajamani 2014).

debated aspect of CBDR-RC, with its spectrum of interpretations ranging from the concept of 'sovereignty', where emissions entitlements shall be given based upon a strict status-quo allocation (Young and Wolf 1992), to that of 'egalitarianism', which implies an equal allocation of emissions for each member (Grubb 1995), plus a wide variety of intermediate options (for a comprehensive review see Berk and den Elzen 2001; Leimbach 2003; Méjean et al. 2015). As a consequence, the long-lasting debate on the interpretation of the CBDR-RC and its components has led to a mix of different, at times diverging, distribution proposals (for an overview, see Höhne et al. 2014; Winkler and Rajamani 2014), none of which were ultimately adopted by the PA.

In adopting CBDR-RC as one of its guiding principles, the PA provided a clearer definition of the concept. By combining elements from different interpretations and by linking them to the need for immediate and robust mitigation action at a global level, the PA suggests an interpretation of 'common responsibilities' as 'responsibilities for all'. This implies that none of the parties involved can be exempted *a priori* from actively participating in the efforts of containing climate change (Article 3 of the PA). Simultaneously, the PA repeats that 'differentiated responsibilities and respective capabilities' must be taken into account in determining the efforts expected from the different parties (as specified in Paragraph 2 of Article 2 and confirmed by the wording adopted in Paragraph 4 of Article 4 and Paragraph 3 of Article 9 of the PA).

While this perspective adopted by the PA appears to be a satisfactory compromise and a substantial contribution to the legal definition of CBDR-RC, its implementation system fails to meet this perspective. While still invoking the relevance of elements such as responsibility and capability, typically used in an egalitarian approach to burden sharing, the PA adopted an (I)NDC-based pledge-and-review method for a self-definition of responsibilities, which instead represents to some extent the supremacy of a sovereign-based perspective. By excluding *de facto* any of the burden sharing approaches proposed by the literature, the implementation system of the PA has already shown its limits in terms of effectiveness and equity (Peters et al. 2015; du Pont et al. 2016; Rogelj et al. 2016).

Once more, A6PA provides an opportunity to overcome the current limitations associated with other parts of the PA. By calling for the need 'to allow for higher ambition in their mitigation and adaptation actions and to promote sustainable development and environmental integrity' (Paragraph 1 of A6PA) for those parties that voluntarily decide to cooperate within its framework, A6PA implicitly provides for them the opportunity to experiment with burden sharing systems other than those specified in the PA. Acting on this opportunity, the MA proposal includes the implementation of a unique burden sharing system to be followed by the members of the MA.

The solution adopted by the MA proposal, involving a mathematical formula to assign the shares of mitigation burden or quotas[10] among MA members, is intended

[10] Quota is used in this Chapter and more in general in this book as synonym of 'assigned share of mitigation burden'.

to align itself to the PA and to A6PA, thus meeting the key criteria of the CDBR-RC principle, while also satisfying the principles of transparency and environmental integrity specified in Paragraph 2 of A6PA. In doing so, it draws upon those streams of equity literature that obtained recognition within the PA wording, as well as on some of their related burden sharing findings. The formula is introduced and compared to a synthesis of these streams and findings in order to demonstrate its full adherence to the specified principles.

It is important to stress that the proposed formula does not aim at distributing emissions allowances, a concept typically addressed by other proposals (Nabel et al. 2011; Höhne et al. 2014; Wei et al. 2014; Méjean et al. 2015). It instead distributes a defined amount of mitigation quotas.[11] These quotas do not refer to shares of the individual MA members' carbon emissions but rather represent shares of a single, collective, common goal. Such quotas thus relate to the members' relative 'weight' within the MA aggregate emissions.

Designed to be applied at the beginning of each timetable to distribute the corresponding shares of the target, the distribution model implies initial discrimination among members before the application of the formula proper. Having calculated the per capita level of carbon emissions of each MA member, as well as the MA's aggregate per capita emissions, the resulting values will be used to establish which members will be responsible for which share of the mitigation target corresponding to the given timetable. Only members whose per capita emissions are higher than the MA aggregate will be responsible for achieving the corresponding quotas. All other members will be exempt from quotas during this timetable. In other words, with *PCA* corresponding to the per capita level of carbon emissions of the MA at the beginning of the timetable *T1*, and *PCx* corresponding to the per capita level of carbon emissions of MA member x at that same time: if $PCA > PCx$, then x is exempt from any quota for the duration of *T1*.

It is important to underline that, in order to prevent MA members and corresponding stakeholders from outsourcing their emissions by moving production of highly emitting goods beyond the MA domain and consequently also to prevent any form of carbon leakage, total emissions of each MA member shall be calculated by adding up its internal production of carbon emissions with the production of carbon emissions originated by goods and services imported from non-MA areas. By solving the long-lasting debate on the prevalence of production-based carbon levels over consumption-carbon levels in the registries of emissions (Peters and Hertwich 2008; Droege 2011), this Border Carbon Adjustment (BCA) policy[12] eradicates the carbon leakage issue from the MA proposal, hence substantially supporting its environmental integrity.

The rationale behind the exemption of lower emissions per capita members and the further use of per capita within the formula is linked to the positive correlation between per capita emissions and development across countries and over time as documented by the Human Development Index (HDI). The correlation implies that

---

[11] For a detailed analysis of emissions allowances see Chap. 7 of this book.

[12] For a full description of Border Carbon Adjustments see Chap. 8 of this book.

the most disadvantaged areas of the world (those registering low or very low HDI values) also register very low per capita emission levels when compared to the most advanced countries (Costa et al. 2011). This view is further confirmed by the vast literature identifying per capita emissions as a key statistic to define climate-related asymmetries in terms of equity (Sijm et al. 2001; Leimbach 2003; Böhringer and Welsch 2006; Markandya 2011). By adopting this approach, the MA proposal may align well with the 'egalitarian' approach to equity and burden sharing.

Nevertheless, this latter approach often envisages an equal-per-capita allocation of mitigation burdens (Rose et al. 1998), which does not represent the ultimate goal of the formula. The formula instead follows the perspective of Michaelowa et al. (2005) concerning the deepening and graduation of different burdens/responsibilities, echoing both the 'contraction and convergence' (Meyer 2000) and 'common but differentiated convergence' (Höhne et al. 2006) perspectives on burden sharing, both of which propound a long-term transition from *status quo* to equal per capita emissions. With ties to egalitarian and convergence approaches, and by even including some references to sovereignty stances, the MA formula aims to become a reliable synthesis for settling the debate on CBDR-RC.

As a consequence of the above-introduced exemption system, the formula therefore applies only to a limited number of MA members in each timetable, while assigning to others no obligations in relation to carbon emissions. This discrimination identifies MA per capita emissions, or PCA, as the threshold for members' obligations. This discrimination persists through time, but changes in per capita emission levels over time will automatically lead to changes in the set of members with quotas and in their relative quotas' magnitudes in successive timetables.

The effects of the discrimination process described here are clear. Changes in per capita emissions, both at member and on aggregate levels, are likely to modify the distribution of quotas both in quantitative and qualitative terms. Members with no obligations in a given timetable may find themselves quotas in the following timetable, or vice versa. It is also plausible for members to accelerate their mitigation pathways voluntarily. Such choices may be prompted by the wish to counteract the potentially negative effects (i.e. the move from a non-binding to a binding mitigation condition or the increase of quota), or to reach the possible benefits (i.e. the move from a binding to a non-binding mitigation condition or the decrease of quota) linked to capita variations, both in relative and absolute terms, over the course of the timetable.

To complete the definition of responsibilities, a precise formula is now required to distribute the MA's target share per timetable between the identified, responsible members. The model proposed in this book is based upon the following mathematical formula, which defines the next mitigation target $MT_{j,t+1}$ for any member j with per capita emissions $PC_{j,t}$ greater than PCA_t at the time t:

$$MT_{j,t+1} = (C_{t+1} + V_t) \times \frac{(PC_{j,t} - PCA_t)P_{j,t}}{\sum_{i:PC_{i,t} > PCA_t} (PC_{i,t} - PCA_t)P_{i,t}}$$

where $PC_{j,t} = E_{j,t}/P_{j,t}$ is the member's per capita emissions rate (current emissions divided by current population). Analogously, PCA_t is total emissions of the MA divided by the total population of the entire MA. Note that the mitigation target $MT_{j,t+1}$ for the next timetable is set at time t, the end of the previous period.

While the first term of the formula above,($C_{t+1} + V_t$), requires further explanation, the fraction representing the second half of the formula can be understood to be member j's portion of emissions above the global threshold, divided by the sum of all such portions across the identified responsible members. It thus ensures an equitable distribution of the MA's target of ($C_{t+1} + V_t$) among members with higher than average per capita emissions. The equity of the formula is further emphasised by the adopted net value of per capita ($PC_{j,t} - PCA_t$). This specifies that each member j is guaranteed with a minimum level of per capita emissions corresponding to the MA aggregate one.

The target itself consists of two components: (i) the predetermined overall target C_{t+1} matching the MA's agreed emissions trajectory to limit temperature increase; and (ii) a dynamic 'variations' term Vt defined as follows:

$$V_t = \max\left(0, \sum_{i:PC_{i,t} \leq PCA_t} \left(E_{i,t} - E_{t-1}\right)\right)$$

Based upon data collected through the public registries referred to in Article 4 of the PA and on information provided by the global stocktake referred to in Article 14 of the PA, the variations term *Vt* represents a significant element of the overall formula and of the entire MA proposal. Its application in fact guarantees environmental integrity of the model embodied by the MA. In particular *Vt* accumulates any overall emissions growth from the set of MA members lying below the threshold for responsibilities during the previous period, as well as any possible increase in emissions from MA members lying above the threshold.[13] This amount is then automatically reallocated to the entire MA as part of target for the following timetable.

The system therefore allows the MA to maintain its long-term goals while adapting dynamically to growth trajectories of its members. As many of these members represent developing regions of the world, the CBDR-RC concept is embedded in the formula in a flexible manner, while ensuring environmental integrity. Starting with an overall target translated into a multi-period mitigation distribution (represented by *Ct* for each period *t*), the variations term thus allows some shifting of targets across timetables to adapt to economic or technological uncertainties without compromising the MA's overall NZC target.

The per capita emissions threshold, along with the target distribution in quotas provided by the overall formula, would trigger a process of 'contraction and

[13] It significant to stress here how the MA per se does not involve any specified 'individual peak year' for its members, allowing any of them to eventually increasing their emissions, regardless their position towards the threshold. As reported here, any actual increase is meant to be absorbed by the *Vt* term in the following timetable.

convergence' that would characterise the entire period determined as the timeframe for achieving the MA's NZC target. While other approaches in the literature also claim such a reaction, the scenario analysis of Coulon and Stua (2015) suggest that the proposed formula would accelerate the convergence process due to several key features. By simulating the paths of emissions of all countries (assuming a fully global alliance) through to 2050 using several different versions of the formula and assumptions on the mean and volatility of business-as-usual emissions growth, it has been found that each of the following features leads to accelerated convergence across regions:

- The variations term *Vt*: Emission increases by lower per capita countries trigger more rapid future mitigation requirements for all, thereby accelerating convergence.
- The use of a threshold: Higher emitting countries face steeper mitigation trajectories since others benefit from short-term exemptions.
- The use of $PC_{j,t}$ in formula: Above the threshold, the percentage cuts required are highest for the highest per capita emitters.
- Dynamic nature of formula: Static allocation of target trajectories far in advance are suboptimal as they fail to adapt to unforeseen events and contingencies.

A spectrum of related approaches exists for the challenges of distributing mitigation responsibilities or alternatively allocating permits, either with or without a specific quantitative formula provided. Several studies (e.g. Oberheitmann 2010; Welp et al. 2006) extensively review the common categories of approaches, which stress different notions of capability, responsibility and equity when defining distributions, often advocating weighted combinations of these concepts. (Baer et al. 2009; Raupach et al. 2014; Den Elzen et al. 2007) Historical cumulative emissions per capita, GDP per capita and carbon intensity have all been suggested as alternatives to the common measure of emissions per capita, sometimes in conjunction with participation thresholds, multiple stages, or sectoral approaches.

Among concepts approximating the formula presented above, the idea of 'Common But Differentiated Convergence' by Höhne et al. (2006) is noteworthy, with non-Annex I Parties exempt from the convergence scheme until their emissions per capita reach a fixed percentage of the global average. However, the mechanism limits full flexibility of transitions of members in both directions across the threshold and also fails to reconcile the uncertainty produced by such exemptions with long-term absolute emissions targets. More recently, GGCC (2015) proposed a burden sharing formula that also compares a party's per capita emissions with global per capita emissions, although they are averaged over the 20 latest years of time and used to distribute more (or fewer) emissions allowances to those further below (or above) the average. Hence, at one end of the distribution very high emissions caps play the role of members' mitigation exemptions, while at the other end caps could become unrealistically low or indeed negative if a historical track record is poor enough. Finally, Sargl et al. (2016) analyse the Regensburg Formula as a mathematical model for implementing convergence of per capita emission levels, and compare several different weighting approaches.

While related approaches in the literature often share a spirit similar to that of the MA proposal, they do not incorporate the full set of desirable features contained in the formula proposed above. In particular, compensating for previous periods via V_t to ensure the preservation of overall mitigation targets in the face of emissions uncertainty is a crucial innovation. Moreover, this same innovative feature is strongly linked to the notion of equity, since the existence of the dynamic threshold criterion offers members temporary exemption from any mitigation responsibilities in the short term, while ensuring incentives to act to retain this status in the future.

Providing such flexibility and dynamic adaptability is typically challenging in a more traditional burden sharing system that distributes allowances instead of mitigation quotas. Although many variations on the distribution formula can be imagined and should be fully explored and compared, the proposed system provides a strong foundation for a standardized mechanism that brings together the criteria of equity, transparency and efficiency in incentivizing mitigation, while simultaneously addressing the practical challenges of uncertainty, limited foresight and long-term environmental integrity. As a consequence it is possible to affirm that the MA formula is aligned with the PA's CBDR-RC principle.

5.4 Conclusions: Significance of a Mitigation Demand in Line with the Paris Agreement Objectives

Hailed by most commentators as a breakthrough for international climate policy, the PA itself still contains significant uncertainties, if not shortcomings, to its own effective and equitable implementation. The lack of obvious tools capable of guaranteeing alignment between the overall objectives of the PA and the individual actions required to meet them represents by far the biggest threat to the future of the PA.

This chapter demonstrates that solutions capable of overcoming such a threat can be found hidden away within the PA itself. By making it indeed possible for new governance regimes to emerge, A6PA opens the door to ambitious experiments that can effectively eliminate the threat posed by the apparent lack of suitable tools.

Based upon the A6PA framework, the MA proposal offers a promising option in this direction. By adopting a NZC aggregate target aligned with the overall objectives of the PA, as well as by defining a technical, dynamic distribution system for this target over time and among the members of the MA, the proposal fulfils both the effectiveness and the equity requirements of the PA while also guaranteeing full respect of the CBDR-RC principle.

The identified target does not only guarantee that individual MA members' quotas will be aligned with the aggregate objectives of the PA. By clearly defining the quantity of mitigation required to meet this alignment, the target defines also the overall mitigation demand within the MA. To satisfy such demand, it is now necessary to design a mitigation supply system. This is based upon the new *mechanism* introduced in Paragraph 4 of Article 6, and is described in Chap. 6 below.

References

Babiker, M. H., & Eckaus, R. S. (2002). Rethinking the Kyoto emissions targets. *Climatic Change, 54*, 399–414.

Baer, P., Athanasiou, T., Kartha, S., & Kemp-Benedict, E. (2009). Greenhouse development rights: A proposal for a fair global climate treaty. *Ethics, Place and Environment, 12*(3), 267–281.

Berk, M., & den Elzen, M. (2001). Options for differentiation of future commitments in climate policy: How to realise timely participation to meet stringent climate goals? *Climate Policy, 1*, 465–480.

Böhringer, C., & Welsch, H. (2006). Burden sharing in a greenhouse: Egalitarianism and sovereignty reconciled. *Applied Economics, 38*, 981–996.

Bretschger, L. (2016). Equity and the convergence of nationally determined climate policies. *Environmental Economics and Policy Studies*. doi: 10.1007/s10018-016-0161-6.

Clémençon, R. (2016). The two sides of the Paris climate agreement: Dismal failure or historic breakthrough? *Journal of Environment and Development, 25*(1), 3–24.

Costa, L., Rybski, D., & Kropp, J. P. (2011). A human development framework for CO2 reductions. *PloS One, 6*(12), 1–9.

Coulon, M., & Stua, M. (2015, August 21). A dynamic mechanism for transparent and equitable distribution of global emissions reductions. *SSRN Working Paper*.

Deleuil, T. (2012). The common but differentiated responsibilities principle: Changes in continuity after the Durban conference of the parties. *RECIEL, 21*(3), 271–281.

Den Elzen, M., Höhne, N., Brouns, B., Winkler, H., & Ott, H. E. (2007). Differentiation of countries' future commitments in a post-2012 climate regime an assessment of the "south–north dialogue" proposal. *Environmental Science and Policy, 10*, 185–203.

Den Elzen, M. C. J., Admiraal, A., Roelfsema, M., van Soest, H., Hof, A. F., & Forsell, N. (2016). Contribution of the G20 economies to the global impact of the Paris agreement climate proposals. *Climatic Change*. doi: 10.1007/s10584-016-1700-7.

Droege, S. (2011). Using border measures to address carbon flows. *Climate Policy, 11*, 1191–1201.

Du Pont, Y. R., Jeffery, M. L., Gütschow, J., Christoff, P., & Meinshausen, M. (2016). National contributions for decarbonizing the world economy in line with the G7 agreement. *Environmental Research Letters, 11*, 054005.

Garibaldi, J. A. (2014). The economics of boldness: Equity, action, and hope. *Climate Policy, 14*(1), 82–101.

GGCC. (2015). The Carbon budgets that will prevent the average temperature of the Earth's surface from increasing by more than 2 °C when compared with the temperature of the pre-industrial era. *Group Governance Climate Change (GGCC) Proposals Volume 1*, Universitat Politècnica de Catalunya.

Grubb, M. (1995). Seeking fair weather: Ethics and the international debate on climate change. *International Affairs, 71*, 463–496.

Höhne, N., den Elzen, M., & Weiss, M. (2006). Common but differentiated convergence (CDC): A new conceptual approach to long-term climate policy. *Climate Policy, 6*, 181–199.

Höhne, N., den Elzen, M., & Escalante, D. (2014). Regional GHG reduction targets based on effort sharing: A comparison of studies. *Climate Policy, 14*(1), 122–147.

IPCC. (2014). Climate Change 2014: Synthesis Report. Contribution of Working Groups I, II and III to the Fifth Assessment Report of the Intergovernmental Panel on Climate Change [Core Writing Team, R. K. Pachauri, & L. A. Meyer (Eds.)]. IPCC, Geneva.

Leimbach, M. (2003). Equity and carbon emissions trading: A model analysis. *Energy Policy, 31*, 1033–1044.

Luderer, G., DeCian, E., Hourcade, J.-C., Leimbach, M., Waisman, H., & Edenhofer, O. (2012). On the regional distribution of mitigation costs in a global cap-and-trade regime. *Climatic Change, 114*, 59–78.

Markandya, A. (2011). Equity and distributional implications of climate change. *World Development, 39*(6), 1051–1060.

Méjean, A., Lecocq, F., & Mulugetta, Y. (2015). Equity, burden sharing and development pathways: Reframing international climate negotiations. *International Environmental Agreements, 15*, 387–402.

Meyer, A. (2000). *Contraction and convergence: The global solution to climate change*. Cambridge: Schumacher Briefings Paperback.

Michaelowa, A., Butzengeiger, S., & Jung, M. (2005). Graduation and deepening: An ambitious post-2012 climate policy scenario. *International Environmental Agreements, 5*, 25–46.

Millar, R., Allen, M., Togelj, J., & Friedlingstein, P. (2016). The cumulative carbon budget and its implications. *Oxford Review of Economic Policy, 32*(2), 323–342.

Nabel, J. E. M. S., Rogelj, J., Chen, C. M., Markmann, K., Gutzmann, D. J. H., & Meinshausen, M. (2011). Decision support for international climate policy – The PRIMAP emission module. *Environmental Modelling & Software, 26*, 1419–1433.

Oberheitmann, A. (2010). A new post-Kyoto climate regime based on per-capita cumulative CO2-emission rights—Rationale, architecture and quantitative assessment of the implication for the CO2-emissions from China, India and the annex-I countries by 2050. *Mitigation and Adaptation Strategies for Global Change, 15*(2), 137–168.

Ohndorf, M., Blasch, J., & Schubert, R. (2015). Emission budget approaches for burden sharing: Some thoughts from an environmental economics point of view. *Climatic Change, 133*, 385–395.

Peters, G. P., & Hertwich, E. G. (2008). Post-Kyoto greenhouse gas inventories: Production versus consumption. *Climatic Change, 86*, 51–66.

Peters, G.P., Andrew, R.M., Solomon, S., & Friedlingstein, P. (2015). Measuring a fair and ambitious climate agreement using cumulative emissions. *Environmental Research Letters, 10*, 105004.

Rajamani, L. (2016). Ambition and differentiation in the 2015 Paris agreement: Interpretative possibilities and underlying politics. *International and Comparative Law Quarterly, 65*, 493–514.

Raupach, M. R., Davis, S. J., Peters, G. P., Andrew, R. M., Canadell, J. G., Ciais, P., Friedlingstein, P., Jotzo, F., van Vuuren, D. P., & Le Quéré, C. (2014). Sharing a quota on cumulative carbon emissions. *Nature Climate Change, 4*, 873–879.

Rogelj, J., den Elzen, M., Höhne, N., Fransen, T., Fekete, H., Winkler, H., Scheffer, R., Sha, F., Riahi, K., & Meinshausen, M. (2016). Paris agreement climate proposals need a boost to keep warming well below 2 °C. *Nature, 354*, 631–639.

Rose, A., Stevens, B., Edmonds, J., & Wise, M. (1998). International equity and differentiation in global warming policy. *Environmental and Resource Economics, 12*, 25–51.

Sargl, M., Wolfsteiner, A., & Wittmann, G. (2016). The Regensburg Model: Reference values for the (I)NDCs based on converging per capita emissions. *Climate Policy*. doi: dx.doi.org/10.1080/14693062.2016.1176006.

Sijm, J., Janson, J., & Torvanger, A. (2001). Differentiation of mitigation commitments: The multi-sector convergence approach. *Climate Policy, 1*, 481–497.

Tavoni, M., Kriegler, E., Rihai, K., van Vuuren, D. P., Aboumahboub, T., Bowen, A., Calvin, K., Campiglio, E., Kober, T., Jewell, J., Luderer, G., Marangon, G., McCollum, D., van Sluisveld, M., Zimmer, A., & van der Zwaan, B. (2015). Post-2020 climate agreements in the major economies assessed in the light of global models. *Nature Climate Change, 5*, 119–126.

UNFCCC. (1992). United Framework Convention on Climate Change Treaty. Rio 1992.

Van Asselt, H. (2016). International climate change law in a bottom-up world. *Questions of International Law, 26*, 5–15.

Van den Bergh, J. C. J. M. (2016). Rebound policy in the Paris Agreement: Instrument comparison and climate-club revenue offsets. *Climate Policy*. doi: dx.doi.org/10.1080/14693062.2016.1169499.

Waisman, H., Torres Gunfaus, M., Spencer, T., & Marquard, R. (2016, May). Emerging from Paris: Post-2015 process, action and research agenda. *Mitigation Actions Plans and Scenarios (MAPS) Programme*, IDDRI, Paris.

Wei, Y.-M., Wang, L., Liao, H., Wang, K., Murty, T., & Yan, J. (2014). Responsibility accounting in carbon allocation: A global perspective. *Applied Energy, 130*, 122–133.

Welp, M., Wicke, L. & Jaeger C. C. (2006). Climate policy in the coming phases of the Kyoto Process: Targets, instruments and the role of cap and trade schemes. *PIK Report*, 107, Potsdam Institute for Climate Impact Research (PIK).

Winkler, H., & Rajamani, L. (2014). CBDR&RC in a regime applicable to all. *Climate Policy, 14*(1), 102–121.

Young, H. P., & Wolf, A. (1992). Global warming negotiations: Does fairness count? *The Brookings Review, 10*(4), 46–51.

Zaman, P., Rock, N., Hedley, A., & Smokelin, J. (2016). *The Paris Agreement: Leading the pathway to a low carbon economy* (16 Platt's Energy Law Report 4, pp. 131–149), LexisNexis A.S. Pratt.

Chapter 6
A Single Mechanism for the Certification of Mitigation Outcomes

Abstract Carbon mitigation objectives, such as the ones included in the Paris Agreement (PA) and the Mitigation Alliance (MA) proposal, require a recognised system to manage the supply of mitigation outcomes. Experience of the Kyoto Protocol (KP), the Clean Development Mechanism (CDM) in particular, shall serve as the key reference for establishing a new mechanism in line with the post-PA mitigation requirements. Yet, experience by itself may not be enough to guarantee the effectiveness, equity and environmental integrity of the newly developed mechanism. Further innovation is required to solve the issues that afflicted the KP experience and to deal with new challenges posed by the ambitious PA objectives. Article 6 of the PA (A6PA) in general, and its interpretation under the MA proposal in particular, serve as tools for overcoming these barriers. By providing structure, the MA mechanism (MAM) ensures both the supply and certification of mitigation outcomes as required by the new international climate regime established after Paris.

Keywords Mitigation supply • Clean Development Mechanism (CDM) • Mitigation Alliance Mechanism (MAM) • Offsetting • Inclusive Offsetting Scheme (IOS)

6.1 Introduction

Once established an effective and equitable method to determine and distribute its carbon[1] emission mitigation (or simply mitigation) target, the Mitigation Alliance (MA) also needs to provide a strong system of certification to enable its members to meet their assigned share of mitigation burden or quota.[2] Article 6 of the Paris Agreement (PA) provides a clear framework for such a system, especially the *mechanism*[3] introduced in Paragraphs 4 to 7. According to the analysis in Chap. 4,

[1] Carbon Dioxide (CO_2) is the most common Greenhouse Gas (GHG) and is often used as a reference unit to express levels of GHG emission (expressed in CO_2 equivalent, or CO_2eq). In the context of this Chapter and, more in general, of this book the term carbon, expressing CO_2, and the term GHG are synonyms, hence carbon is used as single term to express GHG.

[2] Quota is used in this Chapter and more in general in this book as synonym of 'assigned share of mitigation burden'.

[3] In this Chapter, when evidenced in italic the term mechanism refers explicitly and only to its contextualisation within Article 6 of the Paris Agreement (A6PA).

M. Stua, *From the Paris Agreement to a Low-Carbon Bretton Woods*,
DOI 10.1007/978-3-319-54699-5_6

the mechanism envisaged by Article 6 can and should be applied to any form of mitigation, regardless of origin and the approaches[4] adopted for its resulting outcomes' exchange. Aimed at facilitating and accelerating mitigation, the *mechanism* outlined in Paragraphs 4 to 7 of Article 6 of the PA (henceforth A6PA) suggests itself as a tool for the certification of Internationally Transferred Mitigation Outcomes introduced in A6PA's Paragraph 2, or, based on the MA interpretation discussed in this chapter, Internationally *Transferable* Mitigation Outcomes (ITMOs).

The *mechanism* is not explicitly portrayed as a certification mechanism within Article 6. Yet, its description and, most of all, the notable similarities with its Kyoto Protocol (KP) predecessor, strongly support such an interpretation. Several commentators and policy-makers have highlighted the similarities between the A6PA *mechanism* and the KP's Clean Development Mechanism (CDM). Adopted within the KP as one of its implementation tools,[5] the CDM clearly represents a certification instrument.[6] Once it is understood as the predecessor to the A6PA *mechanism*, the CDM can then be considered a blueprint for the new single certification *mechanism* for ITMOs to be developed under the A6PA framework. Recognising the links between the CDM and the new *mechanism* is key to fully understanding the rationale behind the *mechanism*'s interpretation outlined in this book. Section 6.2 firstly discusses to what extent current literature and policy-makers recognise the close ties between the two mechanisms before outlining point-by-point the most significant similarities between them.

Aside from its certifying role, the second key feature of the MA interpretation of the Article 6 *mechanism* refers to its applicability to any form of mitigation outcome, regardless of origins. This more holistic interpretation of how the *mechanism* may be applied to an unlimited range of outcomes, regardless of origin, provided that they meet MA target, partially clashes with the widely diffused opinion that considers the Article 6 *mechanism* to be an instrument limited to market-based approaches. However, this more holistic interpretation finds its rationale in the detailed analysis of A6PA in Chap. 4 of this book and in a full understanding of the architecture of its predecessor, the CDM. Taken together, these provide sufficient procedural and content elements to support the MA interpretation put forward in this Chapter. Section 6.3 details the key supporting elements whilst introducing the rationale behind the holistic MA approach.

Active since 2006 (but with some projects already approved in 2004 and 2005), the CDM registered more than 8000 projects and programme activities hosted in more than 70 developing countries. Despite the criticism that often accompanied its application, the CDM offers a unique reference for the new mechanism in terms of experience gained. Using the CDM as reference for the new *mechanism* implies an evolutionary interpretation. To start with, it is necessary to explain the advantages of

[4] In the context of Chap. 6, and more in general of this book, the term 'approach' refers to any form of mitigation exchange.

[5] See Article 12 of the KP (UNFCCC 1997).

[6] As confirmed by Paragraph 5 of Article 12 of the KP (UNFCCC 1997).

linking the two mechanisms. Section 6.4 discusses the various elements that confirm the CDM as predecessor of the new *mechanism*, defined Mitigation Alliance Mechanism (MAM) by the MA proposal, by providing a review of CDM application over the years.

The MAM introduced in this book does not represent a re-introduction of the CDM into the post-PA climate regime. It instead signifies a radical innovation based on the lessons learned from its predecessor. Most of the novelties introduced by the MAM contribute to solving issues and critiques that affected CDM implementation over the years. While several relevant changes directly originate from the CDM-related norms included in Article 6 and from the more general principles and objectives of the PA, the MAM also contains certain new and unique elements. Section 6.5 analyses in detail the limits of the CDM and the innovations introduced through the MAM, highlighting their rationale, structure and future significance.

The concluding Sect. 6.6 provides a first analysis of the potential impacts of the MAM as a trigger for higher mitigation ambition and as a tool for sustainable development, equity and transparency. This section also discusses its significance as a tool for determining the supply of mitigation outcomes required by the demand originated through the aggregate MA target, its temporal distribution, and its corresponding quotas. This conclusion to the chapter also establishes the basic pillars to discuss the dynamics between demand and supply of mitigation outcomes characterising the MA, which are subsequently analysed in detail in Chap. 7.

Aside from a targeted literature review and an analysis of the key official documents on the use of offsetting mechanisms worldwide, this Chapter makes use of the data and information of 3642 CDM projects hosted in China from 2005 to 2012. Collected through the analysis of corresponding Project Design Documents (PDD) submitted to the CDM Executive Board (EB) over the years, the data refer to almost 50% of total CDM activities registered worldwide. The collected data offer a unique overview of the CDM's historical impact as well as a precious resource to help gain insight into potential effects of adopting the MAM.

6.2 Ties Between the Clean Development Mechanism and Article 6

The *mechanism* introduced in Paragraphs 4 to 7 of A6PA is not explicitly defined as a 'certification mechanism'. Instead it is portrayed as a tool to facilitate and accelerate mitigation for entities that decide to voluntarily collaborate under its framework. Yet, following the logic applied in designing the MA proposal[7] and, more importantly, noting the similarities between the A6PA *mechanism* and the CDM[8] permits

[7] See Chap. 4 of this book.

[8] Explicitly expressed as certification mechanism in Paragraph 5 of Article 12 of the KP.

a certificate-oriented interpretation for the *mechanism* (henceforth MAM), as developed in these pages.

To affirm the certification value of the new *mechanism* introduced by A6PA in its interpretation offered by the MA proposal, it is necessary for this latter demonstrate the close ties linking its mechanism (the MAM founded on the principles of the A6PA *mechanism*) and the CDM. Firstly, it is important to illustrate the wide agreement reached by both scholars and policy-makers concerning these ties ever since the very first studies on A6PA were published. For instance, Marcu (2016) analysis of the *mechanism* introduced in A6PA traces its emergence back to a Brazilian submission to the United Framework Conference on Climate Change (UNFCCC) in November 2014. This particular Brazilian proposal's view on a mechanism to be adopted within the (future) PA stated that: 'The Economic Mechanism shall be comprised of general guidelines related to an emission trading system and an enhanced Clean Development Mechanism (CDM+)' (Brazil 2014, p. 11).[9]

From a policy-making perspective the link between the CDM and the PA has been made repeatedly. Christiana Figueres, at that time Executive Secretary of the UNFCCC, stated that 'experience gained with the Kyoto Protocol's Clean Development Mechanism (CDM) will be critical to the success of the new universal climate change agreement adopted with much excitement in Paris last year' (UNFCCC 2016). Niclas Svenningsen, another prominent representative of the UNFCCC staff, affirmed that 'it is clear that new market approaches and mechanisms under the Paris Agreement can greatly benefit from the experience of the Kyoto Protocol's Clean Development Mechanism (CDM), which has given rise to more than 8000 clean tech projects world-wide' (CPLC 2016).

By analysing some prominent scholars' comments, the close ties linking the CDM to the newly established mechanism within A6PA become even more obvious. For instance, the first 2016 issue of the Carbon Mechanism Review, edited by the Wuppertal Institute for Climate, Environment and Energy in March, widely discussed these ties and included an analysis of the role of lessons learnt from CDM for the proper implementation of A6PA (Raeschke-Kessler 2016) as well as a study about the transition from the CDM to the newly established *mechanism* (Michaelowa and Hoch 2016). A publication by the International Centre for Integrated Assessment and Sustainable Development further emphasises the significance of the CDM experience for the implementation of A6PA (ICIS 2016) by highlighting the links between the two, while a Policy Brief by the Stockholm Environment Institute affirms that 'Article 6.4 is commonly understood to establish a new crediting *mechanism* under UNFCCC authority, with provisions that strongly resemble those of the CDM' (SEI 2016, p. 2).

Several commentators have expressed similar opinions over the months following the PA approval (Mansell 2016; Sharma 2016; Wiseman 2016; Zwick 2016), thus further supporting the interpretation of a strong link between the A6PA *mechanism* and its predecessor. Finally, the World Bank provides an additional, implicit

[9] It is interesting to note that, by coincidence, the CDM also found its origin in a Brazilian proposal submitted to the UNFCC in 1997 (Cole 2010).

support for this interpretation by giving significant space to the present and future of the CDM in its recent report on carbon pricing (World Bank 2016).

These high profile comments and analyses reinforce the close links between the new A6PA *mechanism* and the CDM. A direct comparison of the original texts concerning the two mechanisms[10] can offer an additional insight into these links and provide useful hints for the development of the MAM, based upon the interpretation MA of the A6PA's *mechanism*. Comparing the official texts necessitates a focus on the main differences between them, before analysing their similarities. Three key differences are of importance in this context. Firstly, the CDM-related text occupies an entire Article of the KP, whereas the *mechanism* in the PA is outlined in only four out of nine Paragraphs composing A6PA. The second key difference, already emphasised in this Chapter, is the presence of the 'certification' concept in the CDM (Paragraph 5) and its absence in A6PA. The final, possibly most relevant, difference between the two mechanisms relates to their width of application. The CDM is limited to developing countries (so–called Non-Annex I parties) while the new mechanism established in A6PA is applicable to any entity voluntarily participating in A6PA, regardless of their socio-economic capabilities.

The common understanding that the *mechanism* in Article 6 of the PA resembles both the CDM and the Joint Implementation (JI) mechanism of the KP nevertheless counterbalances the last difference listed above. As the JI allowed the exchange of mitigation certificates between developed countries (so-called Annex I parties),[11] whereas the CDM aimed to facilitate the exchange of mitigation outcomes between developed and developing countries, their combination might represent a form of 'CDM applicable to all'. Several commentators affirmed that, due to the new legal status of parties within the PA, A6PA represents a combination of the CDM and the JI, previously differentiated in the KP, consequently resolving, at least in part, the main differences between the A6PA (*mechanism* on a side and the mechanisms previewed by the KP on the other (ICIS 2016; Marcu 2016; Michaelowa and Hoch 2016; Raeschke-Kessler 2016; Schneider and La Hoz Theuer 2016; Weischer et al. 2016).

While the significance of the above-listed three key differences between the CDM and the A6PA's new *mechanism* (and consequently the MAM) is widely discussed in other sections of this Chapter, the present section focuses on their most explicit similarities. Composed of 10 Paragraphs, Article 12 of the KP (A12KP) shares a similar length with A6PA (composed of 9 Paragraphs). A12KP almost doubles the space given to the CDM in comparison to the space given by A6PA to its mechanism, specifically addressessed in only four Paragraphs. Yet, when carefully analysed, A12KP not only shares elements with Paragraphs 4 to 7 of A6PA, but links to the entire Article 6 text, therefore extending the comparison opportunities

[10] Article 12 of the KP for the CDM and Paragraphs 4 to 5 of A6PA for the new *mechanism*.

[11] The JI is defined by Article 6 of the KP, stating that 'any Party included in Annex I may transfer to, or acquire from, any other such Party emission reduction units resulting from projects aimed at reducing anthropogenic emissions by sources or enhancing anthropogenic removals by sinks of greenhouse gases in any sector of the economy' (KP, Article 6, Paragraph 1).

between the two Articles. For instance, when affirming that the CDM aims 'shall be to assist parties (…) in achieving sustainable development and in contributing to the ultimate objective of the Convention, and (…) in achieving compliance with their quantified emissions limitation and reduction commitments' Paragraph 2 of A12KP appears to share the same aims reaffirmed in Paragraph 1 of A6PA. Similarly, point (a) of Paragraph 5 of A12KP refers to the 'voluntary participation' of the parties to the CDM, which once again recalls the core principles described in Paragraph 1 of A6PA. In stating the 'objective of ensuring transparency, efficiency and accountability through independent auditing and verification of project activities', Paragraph 7 of A12KP appears to share A6PA's principles expressed in its Paragraph 2, while an open interpretation of Paragraph 10 of the A12KP can lead to identify similarities with the introduced in the second Paragraph of A6PA, as to what concerns the use of mechanisms to assist parties in 'achieving compliance in the first commitment period' (Paragraph 10 of A12KP) and 'engaging on a voluntary basis in cooperative approaches that involve the use of internationally transferred mitigation outcomes towards nationally determined contributions' (Paragraph 2 of A6PA).

By analysing Paragraph 4 of A6PA more closely, it is possible to state that it is closely linked to A12KP's Paragraphs (referring to the benefits first specified in the CDM), 5(b) (concerning the measurable and long-term mitigation benefits) and 9 (focusing on the participation of stakeholders other than parties). By focusing on the additionality issue, Paragraph 5(c) of A12KP appears to share, if not the same structure, then at least the same principles of A6PA's Paragraph 5. Perhaps by simple coincidence, Paragraphs 7 in both Articles refer to the role played by the COP/CMA in determining the modalities and procedures for the two mechanisms, while Paragraph 6 of A6PA (concerning the share of proceeds) is almost identical to Paragraph 8 of A12KP.

Now that the close ties between the CDM and the newly established Article 6 mechanism have been established, it is possible to reinterpret the latter in light of the former in order to facilitate the design of the MAM proposal. Moreover, the already established links between the CDM and the A6PA *mechanism* help to further justify the adoption of two key elements for the structure and functioning of this latter (and thereof of the MAM), namely its use as a tool to certify mitigation outcomes emerging from A6PA voluntary cooperation (and thereof from the MA) and its potential use in certifying any form of mitigation, regardless of origins.

6.3 Introducing the Mitigation Alliance Mechanism

Understanding the establishment of a single mechanism to certify any mitigation outcome to meet the MA target, which corresponds to a Net Zero Carbon (NZC) emissions level,[12] requires the clarification of a set of issues that can be summarised in two questions:

[12] See also Chap. 5 of this book.

1. What is the MAM?
2. How does the MAM work?

The following provides a first overview of the founding aspects of the MAM.

1. What Is the MAM? To be consistent with A6PA criteria concerning certification of mitigation outcomes, the MA establishes a single mechanism (the MAM) to certify any mitigation outcome originated through the voluntary cooperation between its members, regardless of origin and final use. Both the ties between the A6PA's aims and principles[13] and the idea of a single certification *mechanism* are based on the implicit link between Paragraphs 2 and 4 of A6PA reviewed under a 'CDM perspective'. Although some observers identified by Michaelowa and Hoch (2016) still see competition between the *mechanism* introduced in Paragraphs 4 to 7 of A6PA and the ITMO approach described in A6PA's Paragraphs 2 and 3, this view appears unfounded. It assumes that the A6PA *mechanism* refers only to market-based exchanges of mitigation outcomes and that the ITMO concept is a tool linked only to direct, bilateral transactions between parties. However, both the considerations regarding the 'unity' of A6PA introduced in Chap. 4 of this book and the most recent interpretations of the ITMO, which describe it as an evolution of CDM-related Certified Emission Reductions (CERs), contradict this perspective (IETA 2016; Marcu 2016; Weischer et al. 2016).

CERs, Defined in Paragraph 3 of A12KP, represent the final outcome of the CDM certification. They are the means of exchange between parties to meet the CDM aims. To better understand how CERs may serve to support the interpretation of unity behind the MAM, it is useful to note how neither A12KP nor A6PA explicitly refer to carbon market approaches when describing their respective mechanisms. The widespread interpretation of the CDM as a 'market-based mechanism' is therefore incorrect, as confirmed by data resulting from the analysis of the 3642 projects hosted by China between 2005 and 2012. For instance, some interesting clues can be detected by looking at the first purchasers of CERs who directly bought them from primary resources, thus avoiding secondary market purchases. While private foreign firms maintained the leading role as CDM direct purchasers, public institutions also had an important role to play. Even excluding direct interventions of state-owned companies such as energy utilities, who were responsible for a significant share of direct purchases by foreign public authorities in the Chinese CDM reality, six government-related institutions and two international organisations were involved in direct purchasing of CERs from China (Table 6.1 describes the authorities and the corresponding number of projects).

Although their number and relevance may appear somewhat limited in comparison to the overall quantity of CDM projects hosted in China, the projects involving direct participation of public foreign authorities in purchasing Chinese CERs do nonetheless play a significant role in the context of this analysis. For instance, they demonstrate that the CDM system could be and was applied not only in market-based approaches, but also in approaches similar to the bilateral ones described in

[13] See Chap. 4 of this book.

Table 6.1 Public purchasers of CERs from CDM projects in China (2005–2012)

Authority	Number of projects
Belgian Ministry of Environment	5
Danish Ministry of Climate and Energy	4
Danish Energy Agency	4
Dutch Ministry of the Environment	5
Norwegian Ministry of Finance	22
Swedish Energy Agency	33
Asian Development Bank	19
World Bank	15

A6PA' Paragraphs 2 and 3. Along with the interpretation of the CDM as predecessor to A6PA *mechanism*, this mix of both public and private approaches appears to confirm the interpretation of the new *mechanism* as a holistic tool capable of encompassing diverse categories of mitigation certifications, therefore aligning to the view provided by the MA proposal.

Following the MA proposal's interpretation of the A6PA *mechanism*, and therefore of its links with the predecessor CDM, it is now possible to propose some significant considerations about the holistic interpretation of the MAM. While explicit considerations concerning the application of the A6PA' *mechanism* (and the MAM) for these specific non-market approaches under the framework introduced in Paragraph 8 of A6PA are not yet evident, A12KP, however, also lacks a direct reference to these non-market approaches. To confirm both the link between the KP mechanisms on a side and the new A6PA mechanism and MAM on the other, as well as to stress their relevance to the non-market approaches outlined above, further stages of analysis and interpretation are required. They are introduced later in this and the following Chapter, although data concerning the use of CDM in China may offer some preliminary clues favouring an inclusive interpretation of the MAM use *vis-a-vis* unspecified non-market approaches outlined in Paragraph 8 of A6PA.

Direct production of mitigation outcomes, where a member of the MA self-produces mitigation as required by its quota,[14] is likely to become a significant reference for non-market approaches. Due to its nature,[15] the CDM could not officially include such options among the approaches covered by its application. Nevertheless, the practice of adopting unilateral-style CDM projects[16] actually diffused in China, accounting for 411 projects. Such unilateral activities may be perceived as an embryonic approach to the self-production of mitigation certificates (Michaelowa 2007).

[14] For further details on direct production see Sect. 7.4 of Chap. 7 of this book.

[15] Which previewed for the final users of CDM's mitigation outcomes to be belong to Annex I parties while stating for CDM's producers to be hosted in Non-Annex I parties.

[16] Projects that did not include any CERs buyer in the PDDs, as explained further below.

Another relevant and emblematic case can be identified in the database of CDM projects approved in China. The case concerns the 27 PDDs referring to projects aimed at constructing methane recovery and utilisation systems for rural households. Such projects can be classified as strongly oriented towards sustainable development where social and environmental considerations precede economic considerations and are therefore ideal candidates for the 'holistic non-market approaches' described in A6PA's Paragraph 8. The PDDs reveal that a majority of projects (20 in total) referred to a single owner and a single CERs buyer. The buyer turned out to be a United Kingdom shell company, registered in Jersey (Channel Islands) on 27 April 2010 under the name of Rural China Development Company. The PDDs specifically included this company as buyer of the CERs and at least eight projects were originally submitted to the EB only 3 days after the company's official registration (on 30 of April 2010). All evidence suggests that Rural China Development Company was a Chinese-held company established for the sole purpose of allowing the methane recovery and utilisation systems projects to fall under the CDM umbrella, therefore representing a peculiar strategy aimed at overcoming the limits imposed by the KP's CDM regulation for CERs purchasing (which stipulated that CERs may only be purchased by private and public stakeholders originally from Annex I countries).

These examples, from the real-life usage of the CDM, support the interpretation laid out in this book of A6PA's *mechanism* (and therefore of the MAM) as a mechanism potentially applicable to any mitigation approach. Such an interpretation of the MAM supports the development of interlinks between Emissions Trading Schemes (ETS) among MA members, possibly leading to the establishment of a single MA Carbon Market (MACM).[17] Yet, the above-developed considerations would also be applicable for the non-market approaches described in Paragraphs 8 and 9 of A6PA, as well as for the bilateral cooperations between parties envisaged in A6PA's Paragraphs 2 and 3.[18]

To sum up, the MAM envisaged by the MA proposal has the potential to be a flexible certification mechanism potentially applicable to any form of mitigation outcome, regardless of origin, occurring within the MA. A final consideration can be made concerning the accessibility to the MAM. In line with the PA's requirement to abide by its objectives and principles, the MAM would be a tool at the disposal of any entity that agrees to voluntarily cooperate indiscriminately under the A6PA framework. Based on these premises the MA envisages equal access and freedom for its members to adopt the MAM, with equal rights to produce, sell and purchase Certified Mitigation Outcomes (CMO).

2. How Does the MAM Work? Answering this raises an interrelated question that follows on from the above discussion: how is it possible to guarantee the necessary flexibility and inclusivity of the MAM? The answer lies in the establishment of a

---

[17] See Chap. 7 of this book.

[18] A detailed analysis of the significance of this holistic approach is provided in Chap. 7 of this book.

hybrid, multi-functional governance model,[19] where different management, use and monitoring tools coexist under a common objective represented by the NZC target. Similar to the MAM itself, the MAM governance proposal draws from both A6PA and A12KP.

In the outline suggesting the framework of the new *mechanism*, A6PA identifies some criteria that help define its functioning. More specifically, Paragraph 4 of A6PA states that the *mechanism* 'shall be supervised by a body designated by the Conference of the Parties serving as the meeting of the Parties to the Paris Agreement',[20] which represents a first, key element of the MAM's governance. Paragraph 4 of A6PA also states that the body has to be designated by the CMA, without specifying whether it has to be integrated as a part of the UNFCCC system. This lack of specification may mark a radical divergence from the governance of the CDM, where the EB in charge of supervising its implementation was fully integrated within the UNFCCC architecture, leading for the body originated by Paragraph 4 of A6PA to eventually be fully independent from the UNFCCC authority.[21]

This point originates from the same logic that characterised the comparison between the A6PA *mechanism* and the CDM architecture. For instance, if the new *mechanism* equally applies to all those parties that voluntarily decide to collaborate under the A6PA framework, the same implicitly excludes those parties that do not wish to voluntarily collaborate under the A6PA framework. While not previewing such discrimination, the CDM involved virtually all parties that signed the KP (some of them as 'CERs buyers' and others as 'project hosts'), therefore making of its EB an integral part of the UNFCCC. In contrast, the discrimination of the new *mechanism*, applied only to those entities that voluntarily decide to cooperate under the framework of A6PA, requires for its *mechanism*'s governing body to only apply to these entities and therefore not being included among the bodies under the auspices of the UNFCCC, which by nature has an insight into the activities of all the UNFCCC members. Therefore, the institution of a different, independent governing body for the new mechanism appears the most logical option. Following this logic, the MA proposal envisages for the body to be designated by the CMA but drafted by the MA's constituting members.[22]

Following this logic, as well as the original function of the EB, the new body shall be in charge of issuing the CMOs, as well as monitoring and verifying their effective implementation. As discussed in Chap. 8 of this book, the newly proposed MA governance also requires a body in charge of the banking of CMOs at the end of each timetable. This might be the responsibility of the body in charge of issuing

[19] For a full description of the MAM and, more in general, of the MA governance model see Chap. 8 of this book.

[20] Paragraph 4 of Article 6 of the PA.

[21] See Chap. 8 of this book.

[22] See also Sects. 8.5 and 8.7 of Chap. 8 of this book.

and monitoring the certifications or, more likely, responsibility may be passed on to a second, complementary body.[23]

Now that its potential basic architecture has been established, the proposed MAM requires additional considerations to complete its overall structure. Assuming that the body/ies take charge of issuing, monitoring and banking credits, the result may be a top-down interpretation of the overall MAM governance model. This view within the context of the MA proposal, however, would be mistaken. To counterbalance top-down governance of the MAM, represented by the abovementioned body/ies, the MA provides high degrees of flexibility for MA members and stakeholders in the application of its MAM. For instance, the holistic approach discussed above guarantees MA members the opportunity to adapt the mechanism to their own particular policies in a variety of areas, including, *inter alia*, climate mitigation and adaptation, energy, finance, manufacturing industries, agriculture, poverty alleviation, environmental improvement. The possibility to adapt the MAM to a wide range of policy and governance frameworks, as well as the different sectors of application, are further discussed in Chaps. 6 and 7 of this book. Its possible effects are discussed in Chaps. 8, 9 and 10 as well as in the conclusions and policy recommendations.

Flexibility and hybrid qualities of the MAM are also evident when looking at the stakeholders that altogether may involved in the MA, such as firms, energy utilities, financial corporations, but also social and environmental movements, communities, local and sectoral public authorities, down to individuals. The lack of differentiation between MA members, as long as they are not differentiated according to specific norms within their own jurisdictions, potentially grants every stakeholder operating within the MA the freedom to use the mechanism to produce, trade and purchase CMOs, thereby providing them with exclusive opportunities.[24] While the potential consequences of this freedom are analysed in more detail in Chaps. 8, 9 and 10 of this book, the following Sect. 6.4 provides a clear definition of the MAM's features that enable MA members and stakeholders to use it. The section describes the CDM features that provided the basis for designing the MAM and Sect. 6.5 discusses the limitations of the CDM and the novel means to overcome them introduced by the MA.

6.4 The Mitigation Alliance Mechanism as an Evolution of the Clean Development Mechanism

Aside from the similarities analysed above, there are several reasons to use the CDM as the natural starting point for the MAM. The CDM was the most widely diffused mitigation accreditation mechanism. It involved virtually all parties that

[23] See Sects. 8.5 and 8.6 of Chap. 8 of this book.

[24] Comparable with the private excludable goods specified in the climate clubs literature analysed in Chap. 3 of this book.

ratified the KP, with developing countries (Non-Annex I parties) benefitting from hosting CDM activities and developed countries (Annex I parties) benefitting from cost-efficient ways of offsetting their quotas. Before discussing in more detail the positive aspects of the CDM, it is important to stress how the adoption of a CDM-style MAM can provide advantages for MA members from a technical perspective.

Firstly, as a consequence of broad participation of countries in the CDM, MA members will already be familiar with its technicalities. MA members previously falling into the Annex I party category will have advantages in terms of certification purchasing capacity, whereas members previously falling into the non-Annex I category will benefit from their accumulated capacity and knowledge in hosting mitigation activities. This second aspect represents a substantial advantage compared to the first category of members as hosting mitigation activities implies adherence to the principle of Common But Differentiated Responsibilities and Respective Capabilities (CBDR-RC). As a matter of fact, members targeted as a result of the CBDR-RC principle (developing and least developed countries and/or sub-national authorities from these areas) will therefore enjoy a substantial advantage if a MAM is adopted with rules similar to the CDM.

Secondly, the experience of implementing and running CDM activities would guarantee fast and smooth integration of the MAM within the proposed MA, thereby facilitating its quick entry into force and its effective implementation. More specifically, the familiarity with the CDM project cycle, which implied a multi-level governance system, represents invaluable experience for the new mechanism. Based upon a five-stage cycle (project design, validation/registration, monitoring, verification and certification, CERs issuance), the CDM entailed participation of four types of stakeholders: (a) private and public stakeholders with the role of project designers (including projects owners, consultancy companies specialised in project design and eventually first CERs purchasers); (b) international institutions in charge of registering projects and issuing CERs (the Executive Board of the CDM) and monitoring the general functioning of the system (the UNFCCC Conference of Parties); (c) national authorities in charge of project approval (the so called Designated National Authorities, present both in hosting countries and eventually in countries of origin of the first CERs purchasers); (d) private stakeholders in charge of validating and verifying CDM projects (so-called Designated Operational Entities – DOE).

While a full description of the architecture and functioning of the CDM cycle can be found in Netto and Barani Schmidt (2009), it is here possible to affirm that, despite the accusations of excessive technicalities (discussed in more detail in Sect. 6.5), the CDM multi-level model facilitated a widespread familiarity with new, hybrid governance models, where top-down and bottom-up stances, as well as market-based and policy-oriented efforts, blend into a single architecture (Lin 2009; Kang and Park 2013; Methmann 2013; Michaelowa and Michaelowa 2016). It therefore provides a solid foundation for the MA proposal.

Thirdly, access to already tested and verified rules, modalities and procedures represents a huge push for accelerating an efficient deployment of the MAM. Indeed, the CDM accumulated a solid store of technical experience 'capital', which may represent a valuable asset in designing the MAM. This experience 'capital' in governing offset mechanisms gained by the CDM and its 'learning-by-doing' implementation process thus represent a valuable background for establishing the technical structure of the MAM (UNFCCC 2016). Well-established rules, in the form of baselines and methodologies, represent the core elements provided by the CDM experience. Originally established on a project-based principle, the baseline system was aimed at guaranteeing an effective and transparent application of the CDM in any sector affected by its activity (Michaelowa et al. 2007). Baselines included both carbon removals by sinks (A/R baselines) and carbon reduction activities (non-A/R baselines), which were used to calculate the impact in terms of carbon mitigation of the different CDM projects and therefore to determine the corresponding CERs.

The UNFCCC originally defined the A/R baselines as 'the scenario that reasonably represents the sum of the changes in carbon stocks in the carbon pools within the project boundary that would occur in the absence of the A/R CDM project activity', while describing the non-AR baselines as 'the scenario for a CDM project activity that reasonably represents the anthropogenic emissions by sources of carbon that would occur in the absence of the proposed CDM project activity' (UNFCCC 2009). It thereby apparently confirms the project-based orientation of the CDM. Yet, thanks to the learning process that characterised the evolution of the CDM, the baseline concept changed and new elements have been included. Baselines to support programmes were identified through the so-called Programme of Activities (PoA) system, which enabled the registration of coordinated policy implementation under the CDM, leading to emission reductions, therefore further supporting sustainable development actions within the system. Designed as the key tool for the Measuring, Reporting and Verification (MRV) process to guarantee accountability of CDM activities, project-specific baseline setting has often been criticised for its excessive technicalities and complexities (Hayashi and Michaelowa 2013). Complex MRV arguably limited access to the CDM in several sectors by failing to ensure environmental integrity (Schneider et al. 2012) and by reducing access for least developed countries (De Lopez et al. 2009). The process also negatively affected projects with highest sustainable co-benefit potential (for a comprehensive analysis of the technical and bureaucratic lock-ins of the CDM see Shishlov 2015).

To overcome the limits posed by project-specific baseline setting (including the limits linked to the additionality principle that are further analysed in Sect. 6.5), the CDM system also opened to standardised baselines over the years. Standardised baselines are defined by the UNFCCC as 'a baseline established for a Party or a group of Parties to facilitate the calculation of emission reductions and removals and/or determination of additionality for clean development mechanism project activities, while providing assistance for assuring environmental integrity" (UNFCCC 2011). Aimed at overcoming limits associated with excessive technicalities of project baselines, soas to facilitate access to the CDM, the standardised baselines shift the focus of carbon and climate finance activities from project level to sectoral level (Arens et al. 2015). Based on UNFCCC decisions (UNFCCC 2010), standardised baselines can be established in two non-mutually excludable ways: performance standards (emissions or energy intensity denominated in emissions or energy per unit of output and determined by a peer group of similar plants or installations) and default values (emissions factors and emission reduction values used as inputs to the calculation of baseline, project, or emissions leakage). With the approval of 31 baselines, the standard-based system possibly represents the CDM's most important and debated reform up to date.

Since its approval by the UNFCCC, the standardised baseline significantly increased interest in the CDM and became the most commonly used baseline to the mechanism (Arens et al. 2015). While some commentators saw in baseline standards the prefect solution for simplifying the CDM process, therefore advocating its binding use (Spalding-Fecher and Michaelowa 2013), some issues were raised about its capability to guarantee transparency of mitigation outcomes, with special reference to the performance-based standards (Schneider et al. 2012; Hayashi and Michaelowa 2013). Due to their effectiveness, standardised baselines represent the preferred option for the here-proposed MAM mechanism, with the limits and related possible solutions further discussed in Sect. 6.5.

CDM methodologies helped regulate the range of actions that could be supported by the Kyoto mechanism and can therefore be used as reference foundation to determine the MAM's future scope. Based upon a proposal system, where new methodologies could be submitted at any time in order to extend the effectiveness of the CDM, the methodological system guaranteed a high degree of flexibility within the whole CDM system. To date, the latter registered 4 methodologies concerning A/R activities and 207 methodologies concerning non-A/R activities, covering a wide range of actions in a variety of sectors, including *inter alia* energy demand, production and efficiency, manufacturing industries, mining, chemicals, agriculture and biomass, waste management, transport, building, fugitive emissions from fuels and reforestation.

Fourthly and finally, the CDM literature provides several insights on area- and sector-specific benefits that can result from the adoption of a mechanism similar to the CDM architecture. While confirming the benefits in terms of cost-effectiveness for stakeholders that decided to offset their quotas (Strand and Rosendahl 2012; Erickson et al. 2014), the literature also provides a great deal of evidence on the benefits which arose from hosting CDM projects. These include, inter alia, benefits in supporting: (a) sustainable development (He et al. 2014; Uddin et al. 2015); (b) diffusion of renewables (Zavodov 2012; Costa Martins et al. 2013; Phillips and Newell 2013; Stua 2013; Lim and Lam 2014; Sreekanth et al. 2014) and clean technologies (Bayer and Urpelainen 2013; Lema and Lema 2013; Gandenberger et al. 2015; Potdar et al. 2016): (c) poverty alleviation (Rive and Rübbelke 2010; Crowe 2013; Wang et al. 2013); (d) environmental co-benefits (Zhang and Wang 2011; Murata et al. 2016); (e) stakeholder engagement (Kang and Park 2013; Mathur et al. 2014; Dong and Holm Olsen 2015; Kuchler and Lövbrand 2016); and (f) leverage of private investments (Bayer et al. 2013; Tatrallyay and Stadelmann 2013; Böhringer et al. 2015; Rahman et al. 2016).

Alongside the discussed benefits, however, the CDM system was also heavily criticised over the years. The MAM aims to overcome these issues by adopting new elements laid out in the PA, specifically those contained within A6PA, and through the unique characteristics of the MA proposal itself.

6.5 Overcoming the Limitations of the Clean Development Mechanism

The debate on the CDM's limits has often been characterised by the apparently contradictory critiques concerning its excessive technicalities that reduce investors' interest on the one side, and its lack of effectiveness in providing environmental benefits on the other (Voigt 2009). More broadly, Christiana Figueres, former Executive Secretary of the UNFCCC, addressed four main shortcomings of the CDM system: inefficient operation, insufficient contribution to sustainable development, weakness of the incentive, and weak governance (Figueres and Streck 2009).

Climate literature focused on mitigation has reiterated and expanded these critiques over the years. By entailing high transaction costs (Michaelowa and Jotzo 2005; Chadwick 2006), excessive technicalities have been identified as barriers to the implementation of activities aimed at effective sustainable development (Dirix et al. 2016), while fostering large projects producing big quantities of CERs (Lövbrand et al. 2009). Concerning the latter, one of the most criticised aspects of the CDM was the inclusion of Hydrofluorocarbons (HFC) abatement among its accepted methodologies. The very low cost of abating this typology of carbon emissions, coupled with their high impact in terms of CERs production and their null effect in terms of sustainable development, was identified as one of the least effective choices within the CDM system (Wara 2008; Böhm et al. 2015). An attempt was launched in 2011 to solve this issue by adopting new methodologies concerning HFCs. Now the issue appears to have been solved by the global deal to limit their use reached within the Montreal Protocol on Ozone Layers in October 2016 (Johnston et al. 2016). HFCs are no longer within the scope of the PA and are consequently automatically excluded from the activities linked to the MAM.

Concerning the limits related to excessive technicalities and bureaucracy of the CDM, the PA in general, and A6PA in particular, offer solutions through several complementary novelties. Firstly, the absence of any *a priori* differentiation between producers and buyers and the related requirement to avoid double-counting of mitigation outcomes resulting from the A6PA *mechanism*[25] represent a radical innovation to facilitate the MRV process aimed at validating the outcomes. Moreover, they represent an important step forward to help overcome the additionality issue which, in the view of several commentators, represented the biggest 'lock-out' for the implementation of small, highly sustainable mitigation activities, with special

[25] Paragraph 4 of Article 6 of the PA.

reference to those in least developed countries (Michaelowa and Jotzo 2005; Chadwick 2006; Lövbrand et al. 2009; Spalding-Fecher and Michaelowa 2013; Carmichael et al. 2016).

Aimed at guaranteeing environmental integrity of CDM projects, additionality is defined by UNFCCC (2009) as 'the effect of the CDM project activity to reduce anthropogenic GHG[26] emissions below the level that would have occurred in the absence of the CDM project activity (in case of non-A/R projects)' or 'the effect of the CDM project activity to increase actual net GHG[27] removals by sinks above the sum of the changes in carbon stocks in the carbon pools within the project boundary that would have occurred in the absence CDM project activity (in case of A/R projects)'.

The additionality concept as proposed by UNFCCC was intended to avoid carbon leakage. In practice, it aimed to limit Annex I parties from benefitting from the purchase of CERs resulting from activities that would have happened even in the absence of CDM support, thereby providing Annex I parties with the opportunity to continue emitting at the expense of global carbon mitigation efforts. Due to its hypothetical nature (additionality refers to a scenario that can only be foreseen but never verified), proving additionality became a source of both high transaction costs for project developers (Michaelowa and Jotzo 2005; Chadwick 2006; Wang and Chen 2010; Rahman and Kirkman 2015) and high uncertainty concerning the effective environmental integrity of projects (Schneider 2009; Grubb et al. 2014).

The novelties introduced by A6PA, and the MA proposal can effectively help overcome the additionality issue, reduce transaction costs, and guarantee the necessary environmental integrity of activities implemented through the MAM. Firstly, the abandonment of the Annex I/Non-Annex I discrimination and the adoption of a nuanced interpretation for CBDR-RC, leading to the inclusion 'responsibility for all' principle within the PA, overcome the current concept of 'offsetting'[28] and eliminates the risks related to carbon leakage within A6PA. As a matter of fact, based upon the analysis provided in this Chapter, any certified mitigation, when not self-produced by the stakeholder directly holding a quota, represents an offset. No more distinction between domestic actions and international offsetting is therefore required.

A second key tool to help overcome the additionality is represented by the A6PA's requirement for double-counting avoidance. By establishing that any mitigation outcome exchanged through the new A6PA *mechanism* (and therefore through the MAM) cannot be claimed by both the outcome's producer and its purchaser, double-counting avoidance virtually excludes any additionality-related issue, hence playing a key role for environmental integrity in the new international climate regime. Double-counting avoidance requires the establishment of public

[26] Or carbon.

[27] Or carbon.

[28] Traditionally meant to specify the mitigation of a carbon emission made in order to compensate for or to offset an emission made elsewhere. For a more complete analysis of the 'offsetting' revolution proposed by the MA see Chap. 7 of this book.

registries for exchanging mitigation outcomes at any level, therefore becoming also a relevant tool for granting transparency to any regime set up through A6PA.

In addition to the novelties introduced by the PA generally and by A6PA in particular, a third and possibly even more relevant tool to overcome the additionality limit comes from the MA itself, specifically through the formula designed to distribute the shares of mitigation responsibility between the MA members.[29] The introduction of the element aimed at compensating any increase in aggregate emissions within the MA during the different timetables, the *Vt* term, *de facto* guarantees that the MA will provide net mitigation aligned with its final NZC target, as well as its intermediate mitigation burden. Therefore, a full and transparent application of the formula introduced in Chap. 5 of this book secures the effective environmental integrity of the MA proposal and overcomes the additionality issue for good.

Overcoming the additionality issue offers the opportunity to radically simplify the overall cycle relating to the new MAM and the management of its CMO by favouring the adoption of standardised baselines for CMOs issuance. The MA envisages the binding use of standardised baselines coupled with a voluntary use of programme- and project-based ones. With the first ones supporting greater participation in the mechanism, thanks to their simplified system, the second ones would then still offer the opportunity, for stakeholders that voluntarily adopting them, to demonstrate a higher effectiveness in mitigation outcomes compared to standardised calculations. For instance, to guarantee transparency and environmental integrity, MAM projects and programmes implemented using *ad hoc* baselines shall require a more accurate and detailed MRV process compared to the use of standardised baselines. Adopting the 'double-standards' system shall in principle facilitate greater participation in the MAM, especially by least developed countries' stakeholders, and support the implementation of more sustainable development oriented activities, therefore overcoming the CDM limits in these areas (Martinez and Bowen 2013; Erickson et al. 2014; Dirix et al. 2016) whilst also supporting the equity and CBDR-RC principles of the PA and integral to the MA.

An additional relevant novelty envisaged by the MA concerns forestry activities under its framework. For instance, the links between A6PA and Article 5 of the PA (A5PA), as envisaged by Marcu (2016), can lead to an application of the MAM to activities aimed at Reducing Emissions From Deforestation and Forest Degradation (REDD+), which now fall under A5PA. The application of MAM to REDD+ activities finds its rationale in several elements. Firstly, despite having been active for years, REDD+ has maintained an unsolved institutional fragmentation concerning its funding architecture (Well and Carrapatoso 2016), with unpredictable funds providing uneven support to REDD+ initiatives and poor adoption of proper offsetting mechanisms. The latter aspect can be traced back to two main components: (a) representing an extensive source of emission mitigation, REDD+ was seen as a potential source of disruption for emerging carbon markets, based upon relatively low mitigation ambitions (Anger et al. 2012); (b) governed through a complex, multi-level system without a specific mandate for an international monitoring

[29] See Sect. 5.3 of Chap. 5 of this book.

authority, REDD+ activities lack common, institutionalised methods for calculating their mitigation impact (Brockhaus et al. 2014).

The new PA context and the opportunities offered by the MA resolve both lock-outs of REDD+ by potentially supporting them through the MAM. Firstly, the lack of differentiation between MA members would extend the adoption of REDD+ initiatives into those geographic areas (such as the former Annex I parties) previously not included among its potential beneficiaries. Secondly, the high mitigation potential of REDD+ perfectly adhere to the ambitious target of the MA,[30] potentially facilitating the MA members' meet with their quotas, hence contributing to higher participation in the MA. Thirdly, the recent introduction of the 'results-based payments for REDD+' concept (Wong et al. 2016), which aims at financing REDD+ initiatives on the basis of their effective mitigation outcomes, appears perfectly in line with the general functioning of the MAM. Moreover, the simplified MRV system for the MAM application can counterbalance the transaction costs resulting from the application of result-based payments, thereby avoiding limitations to widespread REDD+ dissemination.

While the envisaged merging of the REDD+ within the new MAM would serve to exponentially increase forestry activities among the MA members in comparison with the relatively poor results of CDM (Aggarwal 2014; Olsson et al. 2016), the adoption of a multi-level governance, described in more detail in Chap. 8 of this book, would radically stimulate greater stakeholder participation compared to the uneven situation under the CDM (Kuchler 2015). Finally, the uneven impact of CDM as a tool for technology transfer (Sreekanth et al. 2014) finds its solution in the multiple approaches that are proposed by the MA in terms of exchange of CMOs, which are illustrated and discussed in Chap. 7 of this book. Bilateral exchanges directly involving national and/or local authorities are the tools most likely to stimulate widespread technology transfer.[31]

6.6 Conclusions: The Mitigation Alliance Mechanism as Pillar of the Mitigation Alliance Supply System

To sum up, this Chapter offers a radically new and enhanced mechanism, the MAM, aimed at certifying any form of mitigation outcome within the MA. The MAM is capable of providing the structure for a holistic instrument because it is applicable to any mitigation outcome among the MA members, thereby overcoming the KP's Annex I/non-Annex I dichotomy. It merges the mechanisms included in the KP (the CDM and the JI) and adds to them the REDD+ initiative. An additional factor in the MAM's holistic architecture is the inclusion of project-based, programme-structured, baseline-organised, and sectoral methodologies for its application,

---

[30] See Chap. 5 of this book.

[31] See Sect. 7.3 of Chap. 7 of this book.

which radically extends its range of action compared to the CDM. Moreover, due to its flexibility, the MAM encourages mitigation in a multitude of areas, including, *inter alia*, energy demand and supply, energy efficiency, manufacturing industries, mining, fugitive emissions from fossil fuels, transport, urbanisation, agriculture, building and real estate, mining, chemical industries, waste management and forestry (both A/R and REDD+). Finally, the MAM supports any CMOs exchange approach, therefore addressing both market and non-market models.

Within the MA, the MAM therefore represents the only way to certify and issue the mitigation outcomes required from the MA members to meet their quotas, which were analysed in Chap. 5 of this book. By including any form of certified mitigation outcome within the MA, the MAM establishes a unique form of offsetting, which also includes self-produced mitigation outcomes. It is therefore appropriate to define the MAM as an Inclusive Offsetting Scheme (IOS). The IOS specifies the functioning of the supply side of the MA proposal, as much as its capability to meet the net NZC target of the MA, hence to satisfy its demand side. By analysing the different approaches to the exchange of mitigation outcomes as envisaged by the MA, the following Chap. 7 of this book introduces a full disclosure on the demand/supply dynamics proposed by the new governance model represented by the MA.

References

Aggarwal, A. (2014). How sustainable are forestry clean development mechanism projects? A review of the selected projects from India. *Mitigation and Adaptation Strategies for Global Change, 19*(1), 73–91.

Anger, N., Dixon, A., & Livengood, E. (2012). Reduced deforestation and the carbon market: The role of market regulations and future commitments. *Environment and Development Economics, 17*, 269–292.

Arens, C., Hermwille, L., & Burian, M. (2015, July). *Implications of standardized CDM baselines for Least Developed Countries (LDCs) and their use in national 'Measuring, Reporting and Verification (MRV)' systems*. Undwelt Bundesamnt Final Report.

Bayer, P., & Urpelainen, J. (2013). External sources of clean technology: Evidence from the Clean Development Mechanism. *Review of International Organizations, 8*(1), 81–109.

Bayer, P., Marcoux, C., & Urpelainen, J. (2013). Leveraging private capital for climate mitigation: Evidence from the Clean Development Mechanism. *Ecological Economics, 96*, 14–24.

Böhm, S., Brei, V., & Dabhi, S. (2015). EDF Energy's green CSR claims examined: The follies of global carbon commodity chains. *Global Networks, 15*, 87–S107.

Böhringer, C., Rutherford, T. S., & Springmann, M. (2015). Clean-development investments: An incentive-compatible CGE modelling framework. *Environmental and Resource Economics, 60*(4), 633–651.

Brazil. (2014, November 6). *Views of Brazil on the elements of the new Agreement under the Convention applicable to all the Parties*. Brazil submission to the UNFCCC.

Brockhaus, M., Di Gregorio, M., & Maridah, S. (2014). Governing the design of national REDD+: An analysis of the power of agency. *Forest Policy and Economics, 49*, 23–33.

Carmichael, D. G., Lea, K. A., & Balatbat, M. C. A. (2016). The financial additionality and viability of CDM projects allowing for uncertainty. *Environment, Development and Sustainability, 18*(1), 129–141.

Chadwick, B. P. (2006). Transaction costs and the clean development mechanism. *Natural Resources Forum, 30*, 256–271.

Cole, J. C. (2010). Genesis of the CDM: The original policymaking goals of the 1997 Brazilian proposal and their evolution in the Kyoto protocol negotiations into the CDM. *International Environmental Agreements, 12*(1), 41–61.

Costa Martins, D. E., Bernardini Seiffert, M. E., & Dziedzic, M. (2013). The importance of clean development mechanism for small hydro power plants. *Renewable Energy, 60*, 643–647.

CPLC. (2016, September). More than two thirds of Latin American and Caribbean Nations' climate action plans refer to a price on carbon. *CPLC News*.

Crowe, T. L. (2013). The potential of the CDM to deliver pro-poor benefits. *Climate Policy, 13*, 58–79.

De Lopez, T., Tin, P., Iyadomi, K., Santos, S., & McIntosh, B. (2009). Clean development mechanism and least developed countries: Changing the rules for greater participation. *Journal of Environment and Development, 18*(4), 436–452.

Dirix, J., Peeters, W., & Streckx, S. (2016). Is the clean development mechanism delivering benefits to the poorest communities in the developing world? A critical evaluation and proposals for reform. *Environment, Development and Sustainability, 18*(3), 839–855.

Dong, Y., & Holm Olsen, K. (2015). Stakeholder participation in CDM and new climate mitigation mechanisms: China CDM case study. *Climate Policy*. doi:dx.doi.org/10.1080/14693062.2015.1070257.

Erickson, P., Lazarus, M., & Spalding-Fecher, R. (2014). Net climate change mitigation of the clean development mechanism. *Energy Policy, 72*, 146–154.

Figueres, C., & Streck, C. (2009). A post-2012 vision for the clean development mechanism. In D. Freestone & C. Streck (Eds.), *Legal aspects of carbon trading: Kyoto, Copenhagen, and beyond*. New York: Oxford University Press.

Gandenberger, C., Bodenheimer, M., Schleich, J., Orzanna, R., & Macht, L. (2015). Factors driving international technology transfer: Empirical insights from a CDM project survey. *Climate Policy, 16*(8), 1065–1084.

Grubb, M., Hourcade, J-C., & Neuhoff, K. (2014). Cap-and-trade and offsets. In M. Grubb (Ed.), *Planetary economics*. Routledge: Taylor and Francis.

Hayashi, D., & Michaelowa, A. (2013). Standardization of baseline and additionality determination under the CDM. *Climate Policy, 13*(2), 191–209.

He, J., Huang, Y., & Tarp, F. (2014). Has the clean development mechanism assisted sustainable development? *Natural Resources Forum, 38*, 248–260.

ICIS. (2016). Carbon Markets Almanac 2016 Global Developments & Outlook.

IETA. (2016, May). A vision for the market provisions of the Paris Agreement. *IETA Special Papers*.

Johnston, C., Milman, O., & Vidal, J. (2016, October 15). Climate change: Global deal reached to limit use of hydrofluorocarbons. *The Guardian*.

Kang, M. J., & Park, J. (2013). Analysis of the partnership network in the clean development mechanism. *Energy Policy, 52*, 543–553.

Kuchler, M. (2015). Stakeholding as sorting of actors into categories: Implications for civil society participation in the CDM. *International Environmental Agreements*. doi:10.1007/s10784-015-9314-5.

Kuchler, M., & Lövbrand, E. (2016). Simulative governance: On the collaborative narrative of civil society participation in the CDM stakeholder framework. *Environmental Politics, 25*(3), 434–453.

Lema, A., & Lema, R. (2013). Technology transfer in the clean development mechanism: Insights from wind power. *Global Environmental Change, 23*, 301–313.

Lim, X.-L., & Lam, W.-H. (2014). Review on clean development mechanism (CDM) implementation in Malaysia. *Renewable and Sustainable Energy Reviews, 29*, 276–285.

Lin, J. (2009). Private actors in international and domestic emissions trading schemes. In D. Freestone & C. Streck (Eds.), *Legal aspects of carbon trading: Kyoto, Copenhagen, and beyond*. New York: Oxford University Press.

Lövbrand, E., Rindefjäll, T., & Nordqvist, J. (2009). Closing the legitimacy gap in global environmental governance? Lessons from the emerging CDM market. *Global Environmental Politics, 9*(2), 74–100.

Mansell, A. (2016, February 19). What's ahead for carbon markets after COP21? *ICTSD*, BioRes Analysis and news on trade and environment.

Marcu, A. (2016, January). *Carbon market provisions in the Paris Agreement (Article 6)* (CEPS Special Report 128).

Martinez, C. A., & Bowen, J. D. (2013). The ethical challenges of the UN's clean development mechanism. *Journal of Business Ethics, 117*(4), 807–821.

Mathur, V. N., Afionis, S., Paavola, J., Dougill, A. J., & Stringer, L. C. (2014). Experiences of host communities with carbon market projects: Towards multi-level climate justice. *Climate Policy, 14*(1), 42–62.

Methmann, C. (2013). The sky is the limit: Global warming as global governmentality. *European Journal of International Relations, 19*(1), 69–91.

Michaelowa, A. (2007). Unilateral CDM—can developing countries finance generation of greenhouse gas emissions credits on their own? *International Environmental Agreements, 7*, 17–34.

Michaelowa, A., & Hoch, S. (2016). How to transition from the CDM to the sustainable development mechanism under the Paris agreement. *Carbon Mechanisms Review, 1*, 28–31.

Michaelowa, A., & Jotzo, F. (2005). Transaction costs, institutional rigidities and the size of the clean development mechanism. *Energy Policy, 33*, 511–523.

Michaelowa, K., & Michaelowa, A. (2016). The growing influence of the UNFCCC Secretariat on the clean development mechanism. *International Environmental Agreements*. doi:10.1007/s10784-016-9319-8.

Michaelowa, A., Gagnon-Lebrun, F., Hayashi, D., Salgado Flores, R., Crête, P., & Krey, M. (2007). Understanding CDM methodologies. A guidebook to CDM rules and procedures. *Perspectives CC and DEFRA Guidebook.*

Murata, A., Jiang, J., Eto, R., Tokimatsu, K., Okajima, K., & Uchiyama, Y. (2016). Environmental co-benefits of the promotion of renewable power generation in China and India through clean development mechanisms. *Renewable Energy, 87*, 120–129.

Netto, M., & Barani Schmidt, K-U. (2009). The CDM project cycle and the role of the UNFCCC secretariat. In D. Freestone & C. Streck (Eds.). *Legal aspects of carbon trading: Kyoto, Copenhagen, and beyond.* Oxford: Oxford University Press.

Olsson, A., Grönkvist, S., Lind, M., & Yan, J. (2016). The elephant in the room – A comparative study of uncertainties in carbon offsets. *Environmental Science & Policy, 56*, 32–38.

Phillips, J., & Newell, P. (2013). The governance of clean energy in India: The clean development mechanism (CDM) and domestic energy politics. *Energy Policy, 59*, 654–662.

Potdar, A., Singh, A., Unnnikrishnan, S., Naik, N., Naik, M., & Nimkar, I. (2016). Innovation in solid waste management through clean development mechanism in India and other countries. *Process Safety and Environmental Protection, 101*, 160–169.

Raeschke-Kessler, K. (2016). How to make Article 6. 4 of the Paris agreement work. *Carbon Mechanisms Review, 1*, 8–11.

Rahman, S. M., & Kirkman, G. A. (2015). Costs of certified emission reductions under the clean development mechanism of the Kyoto protocol. *Energy Economics, 47*, 129–141.

Rahman, S. M., Dinar, A., & Larson, D. F. (2016). The incidence and extent of the CDM across developing countries. *Environment and Development Economics, 21*, 415–438.

Rive, N., & Rübbelke, D. T. G. (2010). International environmental policy and poverty alleviation. *Review of World Economics, 146*(3), 515–543.

Schneider, L. (2009). Assessing the additionality of CDM projects: Practical experiences and lessons learned. *Climate Policy, 9*, 242–254.

Schneider, L., & La Hoz Theuer, S. (2016). Design options for reliable and thorough Paris market mechanisms. *Carbon Mechanisms Review, 1*, 19–24.

Schneider, L., Broekhoff, D., Fuessler, J., Lazarus, M., Michaelowa, A., & Spalding-Fecher, R. (2012, November 20). Standardized baselines for the CDM – Are we on the right track? *Carbon Market Watch Policy Paper*.

SEI. (2016). Ensuring the environmental integrity of market mechanisms under the Paris Agreement. *SEI Policy Brief.*

Sharma, S. (2016, January 21). Carbon markets firmly back on the agenda. *IISD Commentary*.

Shishlov, I. (2015, March). Trendsetter for projects: The clean development mechanism. In V. Bellassen & N. Stephan (Eds.), *Accounting for carbon: Monitoring, reporting and verifying emissions in the climate economy* (pp. 341–389). Cambridge: Cambridge University Press.

Spalding-Fecher, R., & Michaelowa, A. (2013). Should the use of standardized baselines in the CDM be mandatory? *Climate Policy, 13*, 80–88.

Sreekanth, K. J., Sudarsan, N., & Jayaraj, S. (2014). Clean development mechanism as a solution to the present world energy problems and a new world order: A review. *International Journal of Sustainable Energy, 33*(1), 49–75.

Strand, J., & Rosendahl, K. E. (2012). Global emissions effects of CDM projects with relative baselines. *Resource and Energy Economics, 34*, 533–548.

Stua, M. (2013). Evidence of the clean development mechanism impact on the Chinese electric power system's low-carbon transition. *Energy Policy, 62*, 1309–1319.

Tatrallyay, N., & Stadelmann, M. (2013). Climate change mitigation and international finance: The effectiveness of the clean development mechanism and the global environment facility in India and Brazil. *Mitigation and Adaptation Strategies for Global Change, 18*(7), 903–919.

Uddin, N., Blommerde, M., Taplin, R., & Laurence, D. (2015). Sustainable development outcomes of coal mine methane clean development mechanism projects in China. *Renewable and Sustainable Energy Reviews, 45*, 1–9.

UNFCCC. (1997). *Kyoto Protocol to the United Nations framework convention on climate change*. New York : United Nations.

UNFCCC. (2009). CDM glossary of terms, version 08.

UNFCCC. (2010). *Standardized baselines under the Clean Development Mechanism* (FCCC/TP/2010/4). United Nations.

UNFCCC. (2011). Decision 3/CMP.6. Further guidance relating to the clean development mechanism.

UNFCCC. (2016, May 13). *Lessons from Clean Development Mechanism critical to implementation of Paris Agreement*. UNFCCC Article.

Voigt, C. (2009). Responsibility for the environmental integrity of the CDM: Judicial review of executive board decisions. In D. Freestone & C. Streck (Eds.). *Legal aspects of carbon trading: Kyoto, Copenhagen, and beyond.* Oxford: Oxford University Press.

Wang, Q., & Chen, Y. (2010). Barriers and opportunities of using the clean development mechanism to advance renewable energy development in China. *Renewable and Sustainable Energy Reviews, 14*, 1989–1998.

Wang, C., Zhang, W., Cai, W., & Xie, X. (2013). Employment impacts of CDM projects in China's power sector. *Energy Policy, 59*, 481–491.

Wara, M. (2008). Measuring the Clean Development Mechanism's performance and potential. *UCLA Law Review, 55*, 1759–1803.

Weischer, L., Warland, L., Eckstein, D., Hoch, S., Michaelowa, A., Koehler, M., & Wehner, S. (2016, May). Investing in ambition – Analysis of the financial aspects in (Intended) nationally determined contributions. *Germanwatch and Perspectives Briefing Paper*.

Well, M., & Carrapatoso, A. (2016). REDD+ finance: Policy making in the context of fragmented institutions. *Climate Policy*. doi:dx.doi.org/10.1080/14693062.2016.1202096.

Wiseman, V. (2016, May 16). Mitigation update: COP 21 carbon footprint released, CDM provides lessons learned for Paris Agreement. *IISD News*.

Wong, G., Angelsen, A., Brockhaus, M., Carmenta, R., Duchelle, A., Leonard, S., Luttrell, C., Martius, C., & Wunder, S. (2016, May). *Results-based payments for REDD+- Lessons on finance, performance, and non-carbon benefits* (CIFOR infobrief, 138). Center for International Forestry Research (CIFOR), Bogor.

World Bank. (2016, October). *State and trends of carbon pricing*. World Bank Group Report. Washington, DC: World Bank.

Zavodov, K. (2012). Renewable energy investment and the clean development mechanism. *Energy Policy, 40*, 81–89.

Zhang, J., & Wang, C. (2011). Co-benefits and additionality of the clean development mechanism: An empirical analysis. *Journal of Environmental Economics and Management, 62*, 140–154.

Zwick, S. (2016, January 02). 6 weeks after Paris, countries moving to build carbon markets. *The Huffington Post*.

Chapter 7
Approaches to the Exchange of Mitigation Outcomes

Abstract Climate literature and policy typically differentiate between carbon markets and carbon taxation as the two main approaches to pricing carbon and subsequent 'exchange' of mitigation outcomes. Additional approaches, such as bilateral direct exchanges between parties or climate-related quantitative easing, have been discussed over time. A clear distinction between these approaches has been maintained to 'safeguard' their independence. Yet, the correlated fragmentation of policies addressing climate change has led to high degrees of uncertainty regarding their aggregate efficacy. By introducing a single mechanism to certify mitigation outcomes, the Mitigation Alliance (MA) helps converge previously separated approaches towards a common objective without limiting flexibility. As a consequence, the MA provides its members with a significant variety of policy, strategy and action options concerning the exchange of mitigation outcomes, while guaranteeing harmonisation and consistency with its holistic architecture. Organising the exchange approaches into three macro-typologies (market, non-market and hybrid), the Chapter outlines the MA's multi-level, hybrid governance model.

Keywords Carbon markets • Allowances • Mitigation targets • Offsetting • Inclusive Offsetting Scheme (IOS) • Exchange approaches

7.1 Introduction

Following the introduction of the system of demand[1] and the system of supply[2] for mitigation outcomes, the dynamics that can contribute to the integration of these systems can now be identified. Once more, the Mitigation Alliance (MA) draws upon the pillars defined by Article 6 of the Paris Agreement (A6PA) to identify these dynamics. It is commonly recognised that A6PA includes at least three categories of

[1] The aggregate target and its distribution over time described in Chap. 5 of this book.

[2] The single mitigation outcomes' certification system discussed in Chap. 6 of this book.

M. Stua, *From the Paris Agreement to a Low-Carbon Bretton Woods*,
DOI 10.1007/978-3-319-54699-5_7

exchange approaches[3] to mitigation outcomes: (a) bilateral exchanges; (b) carbon[4] markets; and (c) undefined non-market approaches (IETA 2016a; Marcu 2016a; Pizer 2016; RCCs 2016; Stavins 2016; Tulkens 2016; World Bank 2016; Zwick 2016).

By analysing approaches to the transferability of Certified Mitigation Outcomes (CMO) as envisaged by the MA proposal in detail, this chapter provides a higher degree of clarification compared to the vagueness of A6PA. The analysis goes beyond the three-party logic adopted by scholars to date and reviews the concept of 'approach'. While discussing the architecture of the approaches to mitigation outcomes' exchange dynamics incorporated in the MA governance model, this Chapter offers a first glimpse of mitigation areas that can profit most from the newly defined approaches and their potential benefits and limits. The discussion maintains a non-exclusionary logic, which affirms that MA members and stakeholders can adopt one or more of the discussed approaches simultaneously. Various approaches may be linked through the certification system of the Mitigation Alliance Mechanism (MAM) described in Chap. 6. The significant role of these interlinkages is repeatedly emphasised in this Chapter.

This Chapter is divided into two main parts, the first (Sects. 7.2 and 7.3) focuses on the analysis of approaches currently defined as market approaches and the second (Sects. 7.4 and 7.5) aims at generally reviewing the concept of approaches in light of the MA proposal. Part one of the Chapter describes typologies of existing or proposed carbon markets, from pledge-based Emissions Trading Schemes (ETS) and Voluntary Carbon Markets (VCM) to new hybrid options, and discusses the transfer mechanisms of mitigation outcomes currently used within them. Part one of the Chapter provides relevance to the meaning of 'offsetting' within the current climate literature and policy. In analysing the potential impact of the MA proposal on the described carbon markets and exchange mechanisms, the second part of the Chapter describes the changes that the adoption of the MA proposal may imply for various types of carbon markets and discusses their possible convergence towards a single Mitigation Alliance Carbon Market (MACM) architecture.

The second part of the Chapter provides an analysis of the heterogeneous variety of approaches envisaged by the MA for exchanging mitigation outcomes among its members. While most of these approaches may comply with the framework of the MACM, only some of them imply an explicit monetary exchange. While the latter are strictly market-oriented, those not implying a monetary exchange can be classified as non-market approaches, thereby conforming to the general, holistic approaches mentioned in Paragraph 8 of A6PA.

[3] In the context of Chap. 7, and more in general of this book, the term 'approach' refers to any form of mitigation exchange.

[4] Carbon Dioxide (CO_2) is the most common Greenhouse Gas (GHG) and is often used as a reference unit to express levels of GHG emissions (expressed in CO_2 equivalent, or CO_{2eq}). In the context of this Chapter and, more in general, of this book the term carbon, expressing CO_2, and the term GHG are synonyms, hence carbon is used as single term to express GHG.

Although CMO play a prominent role for the functioning of mitigation outcomes' exchange dynamics in the proposal presented here, non-certified mitigation outcomes continue to feature as part of the MA's general architecture. Section 7.6 analyses the role and impact of non-certified mitigation outcomes on the overall MA design and implementation, and discusses both the rationale for their exclusion from the certification system and their hypothetical use within the MA. Section 7.7 concludes the Chapter with a summary of its findings and introduces aspects of the governance of the approaches analysed.

7.2 Towards a Mitigation Alliance Carbon Market

The implications of A6PA for currently existing and future carbon markets have attracted serious attention among scholars and policy-makers to date (ICIS 2016; IETA 2016a; Kehoane et al. 2016; Mansell 2016; Marcu 2016b; Michaelowa and Hoch 2016; Schneider and La Hoz Theuer 2016; Sharma 2016; World Bank 2016). Despite the absence of explicit references to carbon markets within A6PA, commentators have identified indirect linking elements both in Paragraph 2, which focuses on the exchange of Internationally Transferred Mitigation Outcomes (ITMOs), and in Paragraph 4, which is dedicated to the new mechanism analysed in Chap. 6 of this book (IETA 2016b; Kinley 2016; Marcu 2016a; Schneider et al. 2016; Sharma 2016; Stavins 2016; Weischer et al. 2016).

A6PA may, therefore, have a variety of effects on any form of existing and future carbon markets, including pledge-based ETS, VCM and recently evolving hybrid types. Following a brief introduction to existing and planned carbon market regimes, this section discusses their respective envisaged categories of mitigation outcomes exchange. While outlining the impact of the MA proposal across the listed categories, this section introduces the Inclusive Offsetting Scheme (IOS) concept. Following the holistic approach that characterises the MAM and the whole MA proposal generally, the discussion centres on the capacity of the MA to change the different markets dynamics as well as to establish a single, hybrid model of carbon market, defined here as the Mitigation Alliance Carbon Market (MACM).

Based on cap-and-trade regimes, pledge-based ETS find their origin in the European Union EU-ETS, still considered the largest operative and spreading ETS. Currently existing and planned ETS, which include multinational, national, regional and local models, are based upon compliance systems and managed through allowance allocation systems. They may vary regarding the included carbon emission's mitigation (or simply mitigation) sectors, the methodology of emissions allowances or mitigation targets[5] distribution, the magnitude of targets, as well as stakeholders involved. At present, it is possible to identify eight ETS already functioning (ICIS 2016): three national ETS (Kazakhstan, New Zealand and South

[5] It is here required to distinguish between emissions allowances and mitigation targets. Emissions allowances, which represent the currently most diffused form of ETS implementation, refer to 'rights to emit' under a mandated aggregate cap. Mitigation targets instead represent duties to mitigate.

Korea); one interregional ETS (the Regional Greenhouse Gas Initiative – RGGI), comprising the American states of Connecticut, Delaware, Maine, Maryland, Massachusetts, New Hampshire, New York, Rhode Island and Vermont; one hybrid form of ETS (the Western Climate Initiative – WCI), where local authorities from different countries (the American state of California and the Canadian provinces of British Columbia, Ontario and Québec), have established a direct ETS-linking process; one city-based ETS in Tokyo, Japan; and one multinational ETS (the EU-ETS). The latter has been extended to the three members of the European Economic Area (EEA) and to Switzerland via an ETS-linking process. A ninth example is the currently evolving Chinese ETS, which is now moving from a pilot scheme involving seven local ETS to an effective national ETS architecture. Nine more ETS are on track (IETA 2016c), including six national ETS that shall enter into force before 2025 (Brazil, Chile, Japan, Mexico, South Africa, Turkey) and three after 2025 (India, Indonesia, Thailand).

Pledge-based ETS are traditionally structured as cap-and-trade systems: once their caps have been defined, ETS jurisdictions allocate allowances to participating stakeholders or target sectors (usually industrial operators in various sectors) which entitle them to track and validate actual emissions in accordance with the assigned amount. ETS jurisdictions require allowances to be retired following a predetermined timetable. Allowances can be distributed free of charge or by auction, and stakeholders that exceed their allowed emissions thresholds are obliged to compensate for their excess emissions by purchasing allowances from stakeholders that kept their emissions level below the mandated cap. Stakeholders can exchange their allowances by three means:

(a) privately by moving allowances between operators within a company and across the ETS jurisdiction;
(b) over the counter by using a broker to privately match buyers and sellers;
(c) trading on a spot market for climate-related assets.

To facilitate ETS members in meeting their targets, or caps, some ETS jurisdictions contemplate the adoption of offsetting mechanisms of both domestic and international nature, and/or allow for linking with other ETS jurisdictions.

Voluntary Carbon Markets (VCM) collect all carbon offsets that are transferred free of charge with the intention of surrendering them into a regular, market-oriented carbon market. The emissions reductions concerned are defined as Voluntary Emissions Reductions (VERs). Companies, non-governmental organizations (NGOs) and individuals are typical stakeholders of VCM. Usually, VERs are purchased with the intent to re-sell or retire them in order to meet carbon neutrality targets. Corporate social responsibility, ethics, reputational or environmental claims, as well as pre-compliance are the main drivers for VCM participation (Koester and Hall 2014). Pre-compliance buyers procure offsets speculatively before the starting date of a compliance carbon market, thus hoping to pay a lower price compared to what the same offset may eventually cost after the start of the compliance programme.

No commonly established rules and regulations govern VCM. The calculations and certifications of VERs are carried out in accordance with a variety of privately created standards. On the positive side, VCM serve as a testing ground for new procedures, methodologies and technologies that may later be included in pledge-based markets. On the negative side, it is actually their voluntary nature itself that implicitly places a limit on the overall quantity of mitigation attained by VCM compared to those achieved within pledge-based ETS. In addition, the absence of a common VCM quality control system and fragmentary certification standards may result in VCM producing VERs of questionable quality (Passero 2010; Marcello 2011).

Whilst the Verified Carbon Standard (VCS) comprises the largest share of VCM certifications (Hamrick and Goldstein 2016), the Gold Standard represents the ideal example of synergy between VCM and pledge-based standards. Developed by a group of NGOs with the aim of creating high-quality carbon offset projects within the Clean Development Mechanism (CDM) and in VCM, the Gold Standard is the first independent and to date the most rigorous certification standard (Passero 2010). Apart from the Verified Carbon Standard (VCS), other standards adopted by VCM include the Climate Action Reserve and the American Carbon Registry. VCM also serve as niches for activities that are too small compared to the administrative burden and transaction costs of CDMs or for mitigation initiatives currently not included in compliance schemes (Karhunmaa 2016; Liang et al. 2016).

VCM are currently characterised by high degrees of uncertainty and instability. Aside from the intrinsic limits associated with voluntary approaches, VCM have registered high degrees of variability both in terms of volume and of price of traded emissions reductions over the years (Hamrick and Goldstein 2016). Co-existence of different certification standards (Brockhaus et al. 2014) and transaction costs which vary depending on the type of offsetting sources (Cacho et al. 2013) also led to high price volatility even within the same timetable (Hamrick and Goldstein 2016).

Hybrid forms of carbon markets, combining elements of pledge-based ETS and VCM, have developed in most recent times. Generally experimental, these new carbon markets are still poorly defined, with limited documentation available regarding how they function. Yet increasing focus on and expectations in carbon markets are arousing interest in their potential, which also requires some investigation in this context. Two emblematic samples of these hybrid markets are the Chinese domestic ETS, set to start in 2017, and the Carbon Offsetting and Reduction Scheme for International Aviation (CORSIA), approved by the International Civil Aviation Organization (ICAO) in October 2016 (ICAO 2016). Aiming to address any annual increase in total carbon emissions from international civil aviation beyond 2020, CORSIA represents the first ever pledge-based sectoral carbon market, which combines the private features of VCM with the compliance structure of ETS. Due to its nature, CORSIA is structured as a purely offsetting system, where stakeholders involved can exchange any type of mitigation outcome certified according to the standards laid down by the ICAO. In principle, CORSIA includes emissions allowances from traditional ETS among its offsets, adopting a linking process that requires them to comply with ICAO standards requires allowances to guarantee integrity with ICAO standards.

The Chinese national ETS represents the second example radically of new, hybrid carbon markets. Based on a compliance system, the Chinese ETS differs from other pledge-based ETS regarding its management of mitigation outcomes trading. While still using an allowances distribution approach, China's ETS relies heavily upon a unique form of domestic offsetting (Swartz 2016). Following solid CDM experience (see Chap. 6) and a concluded experimental phase involving seven pilot carbon markets, China incorporated a mechanism that can be interpreted as a national CDM in its national ETS to be launched in 2017. Participation in the mechanism is supposed to be voluntary, thereby including elements from VCM, although both demand and supply of mitigation outcomes are limited to Chinese jurisdiction. The mechanism is designed to produce Chinese Certified Emission Reductions (CCERs) as a substitute for CERs and to apply to a CDM-oriented range of sectors. It is expected that CCERs will become the most significant tool for the success of the Chinese ETS (Lo and Cong 2017). The Chinese ETS will therefore be the world's first ETS that is strongly oriented towards domestic offsetting.

This outline of current types of carbon markets raises the question as to the types of mitigation approaches they favour. A first distinction can be made between allowances exchange and emissions offsetting. The former allows market stakeholders to reassign or trade their allowances, which limits its application to ETS that follow a compliance, allowance-based approach. VCM are therefore virtually excluded. Emissions offsetting approaches are less restrictive than allowances exchange approaches. In its most general meaning, offsetting includes any form of mitigation relocated to offset emissions produced by stakeholders. All carbon markets described in this section include offsetting as a tool to help achieve mitigation targets.

7.3 Interpretation and Use of Offsetting

Various offsetting approaches have been developed over time and their use may vary widely among different types of carbon markets. Traditional offsetting tends to discriminate between two categories (international and domestic offsetting) plus one possible sub-category that may co-exist with any of the previous two (sectoral offsetting), as well as one 'quasi category' represented by ETS linking. The original rationale for International offsetting and ETS linking is to be found in Article 17 of the KP, whereas sectoral and domestic offsetting have not been specifically referred to in the international climate regime. Therefore, international, domestic, sectoral offsetting and ETS linking represent four specific categories of trading for carbon markets and as such are now examined. As part of the analysis of sectoral offsetting, special attention is paid to Reducing Emissions From Deforestation and Forest Degradation (REDD+) as it represents a key case study for the whole category.

The international offsetting approach can be adopted by both pledge-based ETS and VCM. When adopted within compliance market systems, international offsetting implies for ETS stakeholders the opportunity to compensate for their excess emissions in a less costly way than buying allowances from other ETS stakeholders.

International offsetting can therefore promote higher ambitions in the aggregate ETS pledges by reducing the marginal mitigation costs and expanding ETS liquidity (Mehling and Haites 2009; Erickson and Lazarus 2013). However, when applied to cap-and-trade schemes, international offsetting may undermine the price incentive and, on a more general level, the policies linked to ETS adoption (Trotignon 2012) by providing excessive liquidity or by discouraging investments in sectors explicitly targeted by the ETS policies but characterised by high marginal mitigation costs. To date, international offsetting has been widely used in several pledge-based ETS, including Europe (EU), Japan, Switzerland and New Zealand, which authorised international offsetting through the Certified Emissions Reductions (CERs) of the CDM.

Organized as a unilateral exchange system, where CERs could be transformed into ETS allowances but not offsetting units, CDM-related international offsetting suffered greatly from its exclusion by various ETS jurisdictions (Michaelowa 2014). The exclusion of some types of CDM projects and especially of CERs from specific host countries ended up undermining the functioning CDM system. The most important case was the decision by the EU to exclude from its ETS any CERs originated in rapidly developing countries. The decision, which entered into force in January 2013, led to a *de facto* market failure of CDM and the collapse of CER production, while at the same time increasing uncertainty and volatility regarding the EU-ETS system and its corresponding allowance price mechanism (Michaelowa 2014; Stua 2014). Although international offsetting represents only a limited share of the crediting recognised in traditional pledge-based ETS, which are mostly supported by allowances trading, its application is expected to spread widely (e.g. in the Chinese ETS), if not become the dominant crediting unit (e.g. in CORSIA) in the most recently designed compliance systems.

As a consequence of the lack of pledges, international offsetting represents the most typical source of trading within the VCM. The flexibility of VCM, resulting from their private and voluntary nature, often allows stakeholders involved to acquire offsetting certificates from a wider portfolio of mitigation areas, with related investments affecting a greater range of mitigation sectors, compared to those included in pledge-based offsetting (see below the section on sectoral offsetting). With national or local architecture, VCM tend towards domestic offsetting (Hamrick and Goldstein 2016) as a result of locally oriented interests of the stakeholders involved in similar VCM.

Domestic offsetting is a relatively new form of trading for carbon markets. A unique feature of domestic offsetting is the opportunity to offset emissions within the boundary of a single ETS jurisdiction. Within pledge-based ETS, domestic offsetting is mainly used to include sectors not involved in the allowances distribution. Five pledge-based ETS jurisdictions intend to operate domestic offsetting, three of them national (South Korea, Switzerland and China) and two subnational (California and Québec) in nature. Domestic offsetting among members of the EU-ETS as part of the EU Effort Sharing Decision (ESD) system, which represents the European climate programme in areas not included in the EU-ETS, has been suggested in recent years (Borkent et al. 2012; Sandbag 2016), but no agreement on their inclusion has yet been reached. Nevertheless, domestic offsetting appears to be an increasingly promising source of mitigation outcomes thanks to the decision taken by China to include domestic offsetting in its ETS (see above). It also appears to be registering a significant impact on VCM characterised by a specifically local dimension.

Sectoral offsetting represents a crosscutting category that can be applied to both pledge-based ETS and VCM. In addition, sectoral offsetting can be of international and/or domestic nature. The main rationale for sectoral offsetting lies in specific environmental, socio-economic and political strategies aimed at either channelling climate-related investments towards specific sectors (e.g. renewable energies, forestry, industrial efficiency, waste management) or triggering sectoral de-investment processes. When adopted within pledge-based ETS, sectoral offsetting implies the development of ad hoc legislation aimed at specifying the types of offsets accepted within the markets of jurisdictions concerned (Michaelowa 2014). When adopted within VCM, the sectoral distribution depends mainly on the varying cost-effectiveness of the offsetting achieved by individual sectors.

REDD+ represents a significant case study for sectoral offsetting. Although paradoxically excluded from pledge-based markets (Lederer 2011), because of its extensive mitigation potential[6] (Anger et al. 2012; Lubowski and Rose 2013) and because of the lack of a common and institutionalised standard concerning its implementation (Brockhaus et al. 2014; McDermott 2014), REDD+ greatly increased the relevance of sectoral offsetting for VCM. REDD+ represents the majority of forestry-based offsets (Pettenella and Brotto 2012) and the second biggest category of projects (after wind energy) in VCM (Hamrick and Goldstein 2016). Supported solely by VCM-based initiatives and financed through *ad hoc* (public and private) funds, REDD+ mitigation outcomes suffer from the uncertainty that typically characterises VCM systems. Due to their voluntary nature, VCM are unable to provide certainty concerning the financial support for REDD+ initiatives thereby dramatically limiting the guarantees available to secure the implementation of REDD+ long-term strategies (Well and Carrapatoso 2016).

The economic rationale of the ETS linking approach has traditionally been to provide tools for: (a) expanding the size and liquidity of carbon markets; and (b) achieving higher cost-effectiveness in meeting ETS objectives represented by the cap (Blyth and Bosi 2004; Anger 2008; Jaffe et al. 2009). Some scholars add to this initial rationale other factors, such as the tendency of ETS linking to eliminate the distortion of competitive pricing of allowances thanks to the levelling effect it has on the prices of allowances (Tuerk et al. 2009a). They also note the political dimension of ETS linking, which encourages renewed alliances between the jurisdictions of the linked ETS (Ranson and Stavins 2015); this observation echoes ideas from the climate club literature (Keohane et al. 2015).

ETS linking is applied traditionally to pledge-based carbon markets. Two types are commonly recognised: direct versus indirect linking on one hand, unilateral

[6] Lubowski and Rose especially describe how economics literature has identified opportunities for significant cost-effective mitigation in REDD+ (Lubowski and Rose 2013).

versus bilateral/multilateral linking on the other. Their combination can produce four different kinds of ETS linking: direct unilateral; direct bilateral/multilateral; indirect unilateral; indirect bilateral/multilateral. Direct linking is conditional on an explicit decision by at least one (unilateral) of the linked jurisdictions to accept exchanges of allowances from other ETS within its ETS or by two (bilateral) or many (multilateral) linked jurisdictions to accept each other's ETS offsets within their respective ETS, effectively allowing the free trade of allowances among cooperating ETS. Examples of direct, multilateral ETS can be found in the RGGI, the WCI and the EU - EEA linking system. Bilateral direct linking examples include EU-Australia, EU-Switzerland, and New Zealand-Australia systems. At present, no example of direct unilateral linking is registered.

Indirect ETS linking occurs when the same offsetting mechanism is accepted by more than one ETS jurisdiction. Accredited for offsetting by several ETS jurisdictions, the CDM is the mechanism most commonly used for indirect linking. Although ETS follow varying rules in its implementation, those that incorporated the CDM as a compliance tool implicitly adopted a 'shared unit of measurement' which permitted their allowances to be linked indirectly by comparing their respective exchange rates with the CERs (Mehling and Haites 2009; Michaelowa 2014; Ranson and Stavins 2015). Up to now, CDM-based indirect linking has worked exclusively one-way, i.e. where CDM Certified Emissions Reductions (CERs) can be converted into allowances but not vice versa At present, no indirect bilateral/multilateral linking is registered.

Benefit and risk implications of ETS linking have been the source of controversial debate over the years (Rehdanz and Tol 2005; Sterk et al. 2006; Anger 2008; Flachsland et al. 2009; Tuerk et al. 2009b). Loss of control over domestic carbon policies and outsourcing of emissions reduction at low prices have been identified as potentially perverse effects of carbon market linking (Trotignon 2012). Nonetheless, despite persistent uncertainties (Erickson and Lazarus 2013), ETS linking is now widely recognised as an effective instrument to further stimulate ETS development worldwide (Jaffe et al. 2009; Newell et al. 2013; Michaelowa 2014; Ranson and Stavins 2015; Doda and Taschini 2016).

7.4 The Mitigation Alliance Facilitating Convergence Towards a Single Market

This description of carbon markets, covering both pledge-based ETS, VCM and their four trading categories, follows the general classification proposed by Lederer (2012). By contrast, the carbon market system proposed by the MA aims at linking these various systems under a single framework. The MA proposal offers a unique opportunity to reform the structure of the entire carbon market by facilitating the convergence of existing carbon market systems towards a new and single market to be shared by all MA members. More specifically, the MA proposal impacts on carbon markets by: (a) adopting a common system of demand for mitigation outcomes;[7] and (b) establishing a single mechanism to certify the supply of mitigation outcomes in any form.[8] In the following, a separate analysis of the effects of such innovations on already existing and future carbon markets will help to clarify their significance as transformation tools for mitigation outcomes exchange within the MA.

On the basis of a common set of equitable and effective compliance rules, MA membership entails the adoption of a single certification system, the CMO, as the unit of measurement for documenting the assigned share of mitigation burden, or quotas,[9] achievement. By guaranteeing a consistent, minimum demand for CMOs in the short, medium, and long term and thus stimulating the development of a CMOs market, the MA proposal provides an incentive for direct, bilateral interlinking between the MA members' existing carbon markets and the MAM. This results in an indirect, multilateral linking of pledge-based and voluntary carbon markets included within the jurisdictions of MA members. The MA system thus resembles the Emission Reduction Credits (ECRs) model envisaged by Dewees (2001) as opposed to the cap and trade allowance system. MA members and their stakeholders shall be entitled to establish models to convert any of their ETS allowances or offsets into CMOs and vice versa in order to meet the quotas of Net Zero Carbon (NZC) target agreed under the MA or established within their reference markets. Ultimately, two factors implied by the formation of the MA are likely to influence currently existing and new carbon markets: (a) the rules for sharing the NZC target among MA members over time; (b) the need for uniform their certification standards to the CMO ones.[10]

Concerning currently existing carbon markets, the NZC target presents a strong incentive for MA stakeholders to aim for the generation, exchange and collection of CMOs In the short term, high demand for CMOs would accelerate the price convergence process between linked ETS. In the long term, this high demand would facilitate the evolution of different ETS towards a single, common market. The magnitude and specification of the demand for CMOs within the MA would provide interrelated markets' stakeholders with increased budget certainty concerning the price of allowances (Mehling and Haites 2009). In addition and most importantly, the ETS linking process facilitated by the newly designed MAM would considerably accelerate the convergence of the varying prices for allowances and offsets in use among MA members (Flachsland et al. 2009; Tuerk et al. 2009a, b). This last aspect is crucial for the determination of carbon pricing within the MA.[11]

[7] See Chap. 5 of this book.

[8] See Chap. 6 of this book.

[9] Quota is used in this Chapter and more in general in this book as synonym of 'assigned share of mitigation burden'.

[10] CMO standards shall be based upon a combination of experience, gained through the application of CDM and other mechanisms, on transparency and accountability requirements of the A6PA, and on the decisions taken by the MA founding members.

[11] See Chap. 9 of this book.

On the other hand, the MA model entails the adoption of shared standards and norms for the bilateral exchangeability between ETS allowances and CMOs as well as the correlated multilateral linking between the jurisdictions of the ETS involved. Aimed at safeguarding environmental integrity and ensuring the effectiveness and efficiency of the new system, the definition of harmonised standards promises a significant reduction of challenges, most notably in the areas of share distribution, regulatory control, and environmental performance, which have been affecting carbon markets and their instruments to date (Tuerk et al. 2009a). More generally, a single standardised system for the definition of tradable mitigation outcomes, with bilateral functions in respect of interrelated carbon markets and multilateral functions in respect of their linking, will definitely enhance the environmental, economic, social and political benefits that represent the ultimate rationale for carbon markets (Blyth and Bosi 2004; Anger 2008; Jaffe et al. 2009; Tuerk et al. 2009a, b; Ranson and Stavins 2015).

The new standards shall require, *inter alia*, significant limits on the sovereignty of the MA members within their respective ETS, eventually leading to inconsistencies between CMO and specific carbon markets.[12] In the long term, effective standardisation of criteria governing the production of mitigation outcomes in both existing carbon markets and the MAM will facilitate the winding-up process of the former in favour of a single Mitigation Alliance Carbon Market (MACM).[13]

Although an increased bilateral linking between existing pledge-based ETS and VCM with the MAM system is recommended in this Chapter, the MA proposal itself does not make binding this linking. Carbon markets currently operating within the MA members' jurisdictions can freely decide to keep on operating as before with no linkage with the MAM in order to maintain full control of their governing aims and criteria (see also Chap. 8). While not entitled to contribute directly towards meeting members' quotas and towards addressing corresponding obligations, carbon markets based on standards different from those established by the MAM shall still play a significant role within the MA proposal. Their significance is discussed further in Sect. 7.4 that focuses on the impact of non-certified mitigation outcomes on the overall MA architecture.

A consequence of the formation of the MA would be the standardization of its members' separate ETS that would subsequently be linked to the MAM. This in turn would have a significant influence on the design and evolution of future carbon markets among MA members. The need to satisfy new quotas defined by the MA will encourage any new market jurisdiction to align with the MAM standards. It is therefore possible to infer that MAM standards could act as a reference for carbon markets. This would eventually accelerate further both the linking process among carbon markets and their convergence towards a single MACM.

In many respects, the most recent developments in carbon markets, such as the above-described criteria introduced for controlling the Chinese ETS and the CORSIA, appear to address aims and objectives similar to those of the MA proposal

[12] See also Chap. 8 of this book.

[13] See also Chap. 8 of this book.

presented in this book. For instance, studies and discussions about new carbon markets seem to focus on their potential to contribute towards the establishment of both a single carbon market regime and a corresponding single carbon pricing system (ICAO 2016; Swartz 2016). While the establishment of a single carbon market is still perceived as less of a priority for driving mitigation actions compared to national/sub-national ETS (IETA 2016c), the single market option is attracting increasing interest among international institutions such as the World Bank, which has identified in it the potential for establishing a single carbon pricing system (World Bank 2016).

Ultimately, the MA model for a MACM reflects the key features of a global carbon market as identified by Flachsland et al. (2010) The MA model is also consistent with the views of Bodansky et al. (2015) concerning bilateral and multilateral linkages in climate policies. Thanks to its potential as a trigger for linking both existing and future carbon markets (described above), the MAM can generate a process leading to the institution of a MA-wide single carbon pricing. Because CMO are designed as an IOS, and can thus in principle include and harmonize any form of offsetting and/or allowances trading currently active within MA members' jurisdictions, the MA proposal therefore correlates closely with currently evolving carbon markets and carbon pricing (Edenhofer and Kornek 2016; Swartz 2016) and with the perspectives of institutions such as the World Bank (2016). Ultimately, the MA proposal may represent the required trigger for the long-awaited and predicted increase in private and public investments in mitigation actions (Redmond and Convery 2015).

To compensate for the limitations and barriers potentially caused by a high-profile MAM normative and standardisation process in line with the A6PA requirements, the MA proposal provides high degrees of flexibility regarding the ultimate system agreed upon for producing and exchanging CMOs. The MA assumes that both CMOs production and exchange will be based upon monetary transactions. Systems that do not entail monetary transfer are consequently non-market approaches. Both categories of approaches are described in detail in Sect. 7.3. The significant role that non-CMO mitigation outcomes play within the overall MA architecture represents a further compensation for the limits imposed by the norms governing the MAM. This aspect is explained in detail in Sect. 7.4.

7.5 Approaches to the Certified Mitigation Outcomes Exchange

The holistic model envisaged for the MAM for the certification of all forms of mitigation outcome within the MA finds its logical translation in an equally holistic approach to the transferability of outcomes. Suppliers shall be entitled to exchange CMOs in order to allow MA members to meet their quotas. As CMO in principle represent the totality of mitigation outcomes required for meeting the MA members' quotas and the aggregate NZC target, any approach to their transferability shall take place within the MA, thus including both market and non-market based approaches. Despite the necessary use of a 'market-oriented' terminology, the IOS will not just be applied to the market oriented exchanges traditionally envisaged for the exchange of mitigation outcomes, but will also cover types of exchange that fall directly into the broad non-market category.

The following brief digression on the role of MA members in determining the demand/supply dynamics of CMOs leads on to an extended analysis of various approaches to the CMOs exchange system as envisaged by the MA proposal.

As a consequence of the system adopted by the MA for distributing the demand for mitigation outcomes, MA members are both responsible for the collection of and the primary sources of demand for CMOs. Due to MA members' entitlement to distribute the CMOs demand among stakeholders falling inside their jurisdictions, however, these stakeholders can become directly involved in exchange approaches. On the other hand, thanks to the MAM any stakeholder within the MA can potentially supply CMOs to the MA, thereby acquiring a primary role as supplier in the various CMOs exchange approaches.

Nevertheless, MA members can still manage CMOs supply through *ad hoc* regulations and policies, which may alter the natural dynamics governing the interrelated exchange approaches. Finally, MA members can directly introduce norms aimed at regulating the functioning of CMOs exchange dynamics within their jurisdictions, thereby once more altering the natural dynamics of their interrelated approaches. The analysis of the exchange approaches developed in this section does not take into account the above listed variables resulting from the individual MA members' decisions, hence it offers a simple description of various CMOs exchangeability options presented by the MA proposal.[14]

Section 7.3 contains a complete review of the multiple opportunities offered by the MA for CMOs exchange. To describe the variety of approaches for CMOs exchange envisaged by the MA, this section categorises them as follows:

1. market-oriented approaches associated with monetary exchanges;
2. hybrid approaches that may or may not imply monetary exchanges;
3. non-market based approaches which exclude any form of monetary exchange.

The first category comprises approaches traditionally used for exchanges in currently existing carbon markets. They represent archetypal trading systems for mitigation outcomes and/or emissions allowances and can be classified into three (sub-) types of approaches:

(a) transfers through a broker who privately and directly matches suppliers and final users;
(b) trading on a spot market for climate-related assets exchanges;
(c) direct bilateral exchanges between suppliers and final users.

[14] A more detailed discussion of these variables follows in Chap. 8 of this book.

Partially interrelated, types (a) and (b) represent by far the most common approaches in carbon markets to date. The presence of third parties, in form of either brokers or spot markets, lowers transaction costs by reducing asymmetries between purchasers and suppliers and by simplifying market access for suppliers. Entitled to purchase mitigation outcomes for re-sale both to final users and on spot markets, brokers have a key role in the functioning of these two approaches. Spot markets capable of attracting significant quantities of mitigation outcomes increase their stability thanks to their resulting high liquidity and significant predictability regarding carbon prices (Mehling and Haites 2009), thereby potentially increasing their attractiveness for additional mitigation outcomes trading.

Traditionally, most of the mitigation outcomes exchanged in types (a) and (b) have come from sectors and projects guaranteeing significant gains in terms of mitigation outcomes production. This trend is unlikely to change with the MA, as carbon markets shall maintain their original aims and objectives under the new framework. Likewise, stakeholders of the ETS jurisdictions included in the MA will have no reason to change their interests or to abandon their objectives.

Type (c) of the first category represents a rather peculiar approach to mitigation outcomes and its unique features restrict its use in practice, mainly as a result of high transaction costs. The exclusion of third parties requires both suppliers and purchasers to be fully aware of the technicalities surrounding the exchanges of mitigation outcomes. Problems may arise in cases of significant asymmetries between organizations involved, which may lead to distortions in the prices of the exchanged mitigation outcomes, e.g. as a result of large-scale purchasers potentially imposing ceiling prices on small-scale suppliers. As a consequence, the adoption of such an approach is most likely between equals in terms of similar size or similar powers within the MA (e.g. its own members).

These considerations suggest that the direct bilateral approach[15] represents the first of the three approaches commonly envisaged by scholars when discussing A6PA, which concerns bilateral exchanges between parties, while the mediation of brokers[16] and use of spot markets[17] are reflected in the second one (IETA 2016a; Marcu 2016a; Pizer 2016; RCCs 2016; Stavins 2016; Tulkens 2016; World Bank 2016; Zwick 2016), which covers carbon markets.

Direct bilateral exchanges of mitigation outcomes[18] share a second, significant peculiarity. They represent the first hybrid approach to CMOs exchange available in the MA. Whereas the exchange may imply a monetary exchange, it may also be founded on a non-monetary agreement. Technology transfer may become a significant tool to trigger non-monetary agreements, with the cost of technology paid through CMOs exchanged in the transfer. Similar approaches may attract the interest of public and private stakeholders that own technologies and may ultimately encourage strategies to overcome the general profit orientation of markets.

[15] Category 1(c) above.

[16] Category 1(a) above.

[17] Category 1(b) above.

[18] Category 1(c) above.

The various interests that influence these approaches, as well as their effect, are therefore radically different from those of the types of CMOs trading previously analysed. Direct bilateral exchange mainly concerns stakeholders whose primary interests are not exclusively economic. It presents an opportunity for public and private stakeholders to pursue their interests through targeted interventions. Bilateral exchange serves as a governance tool for MA members and other public authorities, while supporting specific interests of private stakeholders.

Taxation represents the second hybrid approach to CMOs exchange available in the MA[19] and can be of four types:

(a) monetary, mitigation-oriented taxation within the MA;
(b) trade sanctions and border tariff adjustments;
(c) CMO-based taxation;
(d) share of proceeds.

Due to their nature, and with the potential exclusion of the fourth type, taxation-based approaches apply solely to public jurisdictions.

While in principle a single MA monetary, mitigation-oriented taxation,[20] similar to the global carbon fee proposed by Jones (2010), is conceivable, its practical feasibility appears limited by the conflicting interests, ideologies and strategies of MA members. Therefore, monetary, mitigation-oriented taxation within the MA shall more likely mean that a public authority, such as a MA member, imposes payments on its stakeholders to finance mitigation activities identified by the public authority. These levies may be imposed anywhere within the MA jurisdiction, thus becoming a potential tool to support public authority interventions. The variety of possible activities that public authorities can support through the adoption of the monetary, mitigation-oriented taxation include, *inter alia*: (a) sustainable development both at public authority and at MA levels; (b) interaction between mitigation, adaptation and resilience outcomes; and (c) specific, cross-cutting strategies to link mitigation to other policies, e.g. welfare initiatives, energy policies, industrial innovation, technological development, security (Andrew et al. 2010; Lim and Kim 2012: Franks et al. 2015). However, effective monetary, mitigation-oriented taxation would also entail significant political and economic costs, mainly due to the periodical adjustments resulting from economic pressures in the involved jurisdictions (van den Bergh 2016). As a final consideration, it is likely for monetary, mitigation-oriented taxation within the MA to be of special interest to local authorities, where the imposition of a monetary levy may lead to direct and easily perceived benefits for those obliged to pay the tax, at least once the tax revenue is efficiently reinvested at local level.

Trade sanctions and border tariff adjustments[21] represent a taxation form strongly supported within the climate club literature (Weischer et al. 2012; Nordhaus 2015; Sabel and Victor 2015; Falkner 2016). Within the MA, trade sanction and border

[19] Category 2 above.

[20] Category 2(a) above.

[21] Category 2(b) above.

tariff adjustments would in principle affect non-MA members, for instance by requiring importers of carbon-intensive products from jurisdictions external to the MA to purchase an amount of CMOs comparable to those required for domestic producers (Keohane et al. 2015), or through more elaborate forms of adjustment (Charnovitz 2014; van den Bergh 2016). Trade sanctions and border tariffs may become significantly effective tools for incentivising non-members to comply with the MA, as well as for discouraging free riding. Despite their significant potential and advantages, trade sanctions and border tariffs are threatened by possible conflicts with already existing international legislation, such as that of the World Trade Organisation – WTO (Cosbey and Mavroidis 2014; Kehoane et al. 2015; Sabel and Victor 2015). Potential conflicts between the WTO and the general architecture of the MA are further discussed in Chap. 8 of this book.

CMO-based taxation[22] would allow competent authorities to impose a fee in kind, based on a direct CMO levy. This approach is likely to diffuse rapidly as a system for distributing the quotas within MA members' jurisdictions, making their stakeholders ultimately accountable for fulfilling the overall obligation. Given the freedom of authorities in designing CMO-based taxation, it may become a substitute for allowance-based models with sectors, industries and individuals bound to more direct mitigation compliance. A non-monetary approach in principle, CMO-based taxation may nevertheless lead to indirect effects implying forms of monetary transfer for the achievement of assigned tax quotas. When not stated otherwise by the authorities governing the taxation system, stakeholders responsible for imposing the levy may, for instance, adopt one of the monetary approaches described above to collect their assigned quotas of CMOs. Typically, the final aim of revenues from CMO-based taxation would be the fulfilment of obligations assigned by the MA to its members, therefore precluding a 're-use' of collected CMOs. Nevertheless, authorities may decide to re-sell part of their tax revenues on the spot market, or use them for bilateral exchanges, in line with their specific policies and strategies concerning climate change and beyond. Due to its non-discriminatory nature, CMO-based taxation would in principle ultimately benefit any form of action producing CMOs. Nevertheless, authorities may adopt specific criteria for the use of CMO-based taxes to address their respective priority policy interests.

'Share of proceeds'[23] is a unique form of taxation within the MA. Already alluded to in Article 12 of the Kyoto Protocol that defines the functioning of the CDM, 'share of proceeds' recurs A6PA. As stated in Paragraph 6, "A share of the proceeds from activities under the mechanism referred to in paragraph 4 of this Article is used to cover administrative expenses as well as to assist developing country Parties that are particularly vulnerable to the adverse effects of climate change to meet the costs of adaptation" (Paragraph 6 of A6PA). As confirmed by the Conference of the Parties serving as the meeting of the Parties to the Paris Agreement (CMA) in accordance with Paragraph 6 of A6PA, this share has specific objectives and, in its most literal interpretation, it shall be managed by the MA body(ies) in

[22] Category 2(c) above.

[23] Category 2d above.

charge of running the MA.[24] Nevertheless, a broader interpretation of this 'share of proceeds' tax could potentially increase its scope of application. For instance, Paragraph 6 of A6PA refers only to a generic share of proceeds to be used for the objectives stated in the text. Under the MA, members and other authorities may eventually add further shares of proceeds aimed at their own, specific objectives. China applied a similar approach to its hosted CDM projects by introducing a discriminative tax on them, based on percentages of shares varying according to the types of sectors involved, e.g. a 2% tax on CERs originating from renewables and up to 65% tax on fugitive chemical emissions (Stua 2013).

'Share of proceeds' taxation may have multiple functions and can be applied at different levels including private operators forcing co-operators to 'devolve' shares of CMOs for specific purposes. On the one hand, similarly to what is defined in A6PA, additional 'share of proceeds' could be used to finance specific actions that would not otherwise have easy access to direct climate funding, such as adaptation and resilience projects and programmes, or to cover administrative expenses incurred by authorities participating in the MA. On the other hand, discriminative 'shares of proceeds' may be used to channel investments towards predetermined sectors. For instance, Chinese discriminative policy relating to the CDM contributed to the prioritisation of renewables, which subsequently became the main target area for CDM projects in the country (Stua 2013). In addition, 'share of proceeds' may be adopted as a tool to finance ad hoc (public and private) funds such as the long debated Green Climate Fund (GCF). While any sectoral impact of 'share of proceeds' taxation would mainly depend on the choices and strategies adopted by the various MA members and authorities, its potential side effects may include distortions of carbon market dynamics. Discriminative applications of 'share of proceeds' may, for example, create variations in both the liquidity and prices of CMO among different markets' jurisdictions, therefore limiting the aggregate effect of the MAM. CMOs collected through 'shares of proceeds' taxation could be used either to implement specific policies and strategies and to pay for administrative costs incurred by the MA, or to meet quotas imposed on entities (MA members and stakeholders) dealing with the same 'share of proceeds' application, thus maintaining both a monetary and a non-monetary feature of this particular approach to CMOs exchange.

An additional option for hybrid approaches to CMOs exchange is the climate-related quantitative easing option, initially developed by France Stratégie as a tool for the European Central Bank to support European climate policies (Aglietta et al. 2015a). Based upon the adoption of mitigation outcomes defined as Climate Remediation Assets (CRA), climate-related quantitative easing focuses on a central bank(s)-led strategy, which redirects traditional quantitative easing manoeuvres towards a more specific mitigation objective. The CRA financial leverage passes through a quantitative easing system that guarantees institutionally supported loans to support the deployment of mitigation actions within the central bank(s) jurisdictions (for a detailed description of the system see Aglietta et al. 2015b). This ECB

[24] See also Chap. 8 of this book.

proposal may be adopted either by entities operating within other clusters of countries, such as the Asian Development Bank or the New Development Bank, and/or by national central banks. When operating under government control, central banks implementing climate-related quantitative easing can choose to support specific sectors or areas of interventions, thereby acting as policy tools for their corresponding jurisdictions. Limitations of such a system lie in the high requirements and associated transaction costs of accountability, transparency and effectiveness on the one hand and the need for flexibility, efficiency, and accessibility on the other.[25]

Direct production and use of CMOs is likely to become the main non-market approach within the MA.[26] MA members and stakeholders obliged to contribute to the fulfilment of quotas are entitled to produce and use CMOs directly as well as to acquire them by means of the approaches listed above. As demonstrating compliance with quotas requires all mitigation to be translated into CMOs, direct mitigating actions must be certified through the MAM if they are to be used to reach their quotas. Therefore, by covering all the steps of the process leading from CMOs production to their final banking/deposit for the purpose of demonstrating compliance with quotas, the offsetting scheme proposed by the MA also potentially includes direct mitigating actions. As a consequence, IOS is conceived as a fully inclusive scheme, designed to meet the needs for achieving MA members' quotas and the aggregate MA target.

Direct production and use of CMOs is likely to become an effective policy tool for authorities aiming to transform macro-sectors (as well as innovation systems) under their auspices, such as energy, manufacturing, forestry, transport or urban planning. On the other hand, direct production and use of CMOs may also be instrumental for much smaller actions carried out by local communities, small firms, and individuals. Aside from REDD+, efficiency in general, and energy efficiency in particular, mitigation actions in these sectors may in fact come to dominate direct production and use of CMOs. Rebound effects typically linked to efficiency actions pose the greatest threat to such approaches (Binswanger 2001; Sorrell 2010; Hanley and Brennan 2012; Rezai et al. 2013). Rebound effects can, for instance, counteract the reduction in per capita emissions gained by the adoption of efficiency initiatives, thus distorting their scale of contribution to the future distribution of quotas among MA members.[27]

Despite the inclusive role of the MAM and its IOS, non-certified mitigation outcomes retain significance within the overall MA architecture. As described in Sect. 7.4, taking proper account of such mitigation types serves, inter alia, to guarantee the environmental integrity of the MA proposal.

---

[25] For further details on climate-related quantitative easing see Sect. 8.7 of Chap. 8 of this book.

[26] Category 3 above.

[27] See Chap. 5 of this book.

7.6 Role of Non-certified Mitigation Outcomes

While in principle the CMO-based approach shall cover the entire portfolio of mitigation outcomes occurring within the MA, non-certified mitigation outcomes may have a role to play outside the MA. In the following analysis of this specific mitigation, attention is drawn to three main aspects:

(a) mitigation that may be excluded from the MAM-based CMO system;
(b) their relevance within the MA architecture;
(c) actions that MA members and other stakeholders may implement to benefit from such mitigation.

There are several reasons why mitigation outcomes may not be certified, including both technical and political causes. Concerning technical causes, standards and regulations that shall apply to the MAM certification system, which aims to safeguard key principles such as environmental integrity and transparency of the mechanism, may de facto become barriers to the inclusion of specific categories of mitigation activities. The CDM experience provides evidence that tight but necessary regulation leads to significant transaction costs for activities concerned, therefore deterring some potential beneficiaries from applying for certification of their actions (Michaelowa and Jotzo 2005; Chadwick 2006). Such lockouts, which mainly affected small-scale projects and stakeholders with limited access to capacity building, may be overcome thanks to the diffused adoption of the standardised baseline system discussed in Chap. 6 of this book. However, it is likely that transaction costs and lockouts will put off some mitigation activities, effectively being from the MAM.

There are various possible political reasons for non-certification of mitigation outcomes. The MA and/or its members may, for instance, exclude some specific sectors or types of mitigation from certification under the MAM. The CDM includes similar discriminatory choices, e.g. by excluding nuclear power and artificial Carbon Capture and Storage (CCS), or through methodological modifications that excluded Hydrofluorocarbons (HFC) capturing and the construction of ultra-supercritical coal-fired power stations. Similar discriminations are mainly driven by environmental and social concerns, perceived as significant threats by the jurisdictions and stakeholders affected. On the other hand, as already anticipated in this Chapter, MA members and other stakeholders may opt to retain or develop approaches for mitigation exchange parallel to and independent from the MAM. Pledge-based ETS and VCM may decide to refuse to link with the CMO-based system in order to keep their full autonomy in defining rules and standards to be applied within their scope. By choosing autonomy, these authorities are free, for instance, to involve sectors excluded by the MAM, or they may simply prefer to decide themselves how best to pursue (part of) their mitigation policies.

Despite being unable to support MA members and stakeholders in meeting their quotas, whose fulfilment requires an explicit use of CMO, non-certified mitigation outcomes of any form and origin are still relevant within the MA's general architecture. In order to guarantee consistency with the PA objectives and principles, the calculation of absolute and per capita emissions of MA members and stakeholders shall be based on public registries maintained by competent bodies under the auspices of the United Framework Convention on Climate Change (UNFCCC). Encompassing any and every variation in emissions levels registered within any area of the world, these registries shall include, *inter alia*, both certified and non-certified mitigation outcomes. Their adoption as references for the obligation distribution system introduced in Chap. 5 of this book means that any registered variation will have an impact on the distribution of obligations. Non-registered mitigation outcomes can still change MA members' absolute and per capita levels of emissions significantly (also in relation to overall absolute and per capita emissions), thus influencing the magnitude of obligations distributed across its members and even their possible exclusion/exclusion from obligation sharing over time. As a consequence, non-certified mitigation outcomes will retain their effective value within the MA architecture.

The binding nature and the magnitude of CMO-oriented demand shall most likely limit the value of non-certified outcomes when compared to the CMOs. Nevertheless, it is possible for MA members and other stakeholder to opt for strategies and actions aimed at stimulating non-certified outcomes. This aspect will become a key tool for maintaining ETS and VCM independence from the MAM, as well as for establishing ad hoc policies to address actions technically or politically excluded from the MAM.

7.7 Conclusions: Converging the Multiplicity

This chapter introduced the wide variety of approaches envisaged by the MA proposal to support the matching of demand and supply of mitigation outcomes. It shows that explicitly market-based, financial, hybrid, and non-financial approaches concerning the demand/supply dynamics for mitigation coexist under the MA. MA members and/or their various stakeholders can apply the approaches discussed as single tools or as combinations, thus ensuring highest degrees of flexibility.

However, the variety of approaches is counterbalanced by the singularity of the two key elements that define demand and supply within the MA. A single technical tool is adopted for the distribution of the MA NZC mitigation target defined *a priori*, while a single mechanism is foreseen for the certification of mitigation outcomes required to document the fulfilment of quotas. These peculiarities of the MA proposal ultimately reduce the number of envisaged approaches by linking them through the IOS concept. Regardless of the approach selected for exchanging mitigation outcomes, MA members and stakeholders shall in principle exchange a single and common certification unit, defined as the CMO. While exchanges of non-certified mitigation outcomes will still be recognised under the MA, CMO possibly represent the key innovation of the entire MA proposal.

Serving as a convergence and linking tool between the various exchange approaches, CMO as their standard unit of measurement act as their common denominator This all implies significant impacts concerning the carbon pricing concept proposed by the MA. It also opens windows of unexplored opportunity for radical monetary innovation. Before reaching Chap. 9, which focuses on these key topics and their potential effects, it is still necessary to complete the discussion on the MA architecture with a description of the MA governance system. Chapter 8 of this book is fully dedicated to this last, fundamental element of the MA architecture.

References

Aglietta, M., Espagne, É., & Perrissin Fabert, B. (2015a, February 24). A proposal to finance low carbon investment in Europe. *France Stratégie Policy Brief.*

Aglietta, M., Hourcade, J.-C., Jaeger, C., & Perrissin Fabert, B. (2015b). Financing transition in an adverse context: Climate finance beyond carbon finance. *International Environmental Agreements, 15*, 403–420.

Andrew, J., Kaidonis, M. A., & Andrew, B. (2010). Carbon tax: Challenging neoliberal solutions to climate change. *Critical Perspectives on Accounting, 21*, 611–618.

Anger, N. (2008). Emissions trading beyond Europe: Linking schemes in a post-Kyoto world. *Energy Economics, 30*, 2028–2049.

Anger, N., Dixon, A., & Livengood, E. (2012). Reduced deforestation and the carbon market: The role of market regulations and future commitments. *Environment and Development Economics, 17*, 269–292.

Binswanger, M. (2001). Technological progress and sustainable development: What about the rebound effect? *Ecological Economics, 36*, 119–132.

Blyth, W., & Bosi, M. (2004). *Linking non-EU domestic emissions trading schemes with the EU emissions trading scheme* (IEA/OECD paper). Paris: IEA/OECD.

Bodansky, D. M., Hoedl, S. A., Metcalf, G. E., & Stavins, R. N. (2015). Facilitating linkage of climate policies through the Paris outcome. *Climate Policy*. doi:10.1080/14693062.2015.1069175.

Borkent, B., O'Keeffe, S., Neelis, M., & Gilbert A. (2012, September 20). *Costs and effectiveness of domestic offset schemes*. Ecofys Final report.

Brockhaus, M., Di Gregorio, M., & Maridah, S. (2014). Governing the design of national REDD+: An analysis of the power of agency. *Forest Policy and Economics, 49*, 23–33.

Cacho, O. J., Lipper, L., & Moss, J. (2013). Transaction costs of carbon offset projects: A comparative study. *Ecological Economics, 88*, 232–243.

Chadwick, B. P. (2006). Transaction costs and the clean development mechanism. *Natural Resources Forum, 30*, 256–271.

Charnovitz, S. (2014, September 29). *Border tax equalization*. Draft paper prepared for the conference on challenges facing the world trade system, Columbia University, Broadway and Johns Hopkins University, Baltimore.

Cosbey, A., & Mavroidis, P. C. (2014). Heavy fuel: Trade and environment in the GATT/WTO case. *RECIEL, 23*, 288–301.

Dewees, D. N. (2001). Emission trading: ERCs or allowances? *Land Economics, 77*(4), 513–526.

Doda, B., & Taschini, L. (2016). *Carbon dating: When is it beneficial to link ETS?* (London School of Economics – LSE working paper). Grantham Research Institute on Climate Change and the Environment, Grantham.

Edenhofer, O., & Kornek, U. (2016). Coordinated CO2 prices and strategic transfers. In R. N. Stavins & R. C. Stowe (Eds.), *The Paris Agreement and beyond: International Climate Change Policy Post-2020*. Cambridge, MA: Harvard Project on Climate Agreements.

Erickson, P. A., & Lazarus, M. (2013). Implications of international GHG offsets on global climate change mitigation. *Climate Policy, 13*(4), 433–450.

Falkner, R. (2016). A minilateral solution for global climate change? On bargaining efficiency, club benefits and international legitimacy. *Perspectives on Politics, 14*(1), 87–101.

Flachsland, C., Marschinski, R., & Edenhofer, O. (2009). To link or not to link: Benefits and disadvantages of linking cap-and-trade systems. *Climate Policy, 9*, 358–372.

Flachsland, C., Marschinski, R., Edenhofer, O., Leimbach, M., & Baumstark, L. (2010). Developing the international carbon market beyond 2012: Options and the costs of delay. In F. Biermann, P. Pattberg, & F. Zelli (Eds.), *Global Climate Governance beyond 2012: Architecture agency and adaptation*. Cambridge, MA: *Cambridge University Press*.

Franks, M., Edenhofer, O., & Lessmann, K. (2015). Why finance ministers favor carbon taxes even if they do not take climate change into account. *Environmental and Resource Economics*. doi:10.1007/s10640-015-9982-1.

Hamrick, K., & Goldstein, A. (2016, May). *Raising ambition: State of the voluntary carbon markets 2016*. Ecosystem Marketplace Report.

Hanley, N., & Brennan, D. (2012). Economics of a low-carbon future. *Earth and Environmental Science Transactions of the Royal Society of Edinburgh, 103*, 149–156.

ICAO. (2016). *On board a sustainable future*. 2016 Environmental Report, ICAO.

ICIS. (2016). Carbon markets almanac 2016 global developments & outlook.

IETA. (2016a, May). *A vision for the market provisions of the Paris Agreement*. IETA Special Papers.

IETA. (2016b). *Bridging the ambition gap – The rise, reach and power of carbon markets*. Greenhouse Gas Market 2016/17 Report.

IETA. (2016c). *GHG market sentiment survey 2016* (11th ed.).

Jaffe, J., Ranson, N., & Stavins, R. N. (2009). Linking tradable permit systems: A key element of emerging international climate policy architecture. *Ecology Law Quarterly, 36*, 789–808.

Jones, D. B. (2010). Proposal for an international carbon fee. *Energy for Sustainable Development, 14*, 245–250.

Karhunmaa, K. (2016). Opening up storylines of co-benefits in voluntary carbon markets: An analysis of household energy technology projects in developing countries. *Energy Research and Social Science, 14*, 71–79.

Kehoane, N., Petsonk, A., & Hanafi, A. (2016). Building a coalition of carbon markets to spur faster, deeper cuts in climate pollution. In R. N. Stavins & R. C. Stowe (Eds.), *The Paris Agreement and beyond: International Climate Change Policy Post-2020*. Cambridge, MA: Harvard Project on Climate Agreements.

Keohane, N., Petsonk, A., & Hanafi, A. (2015). Toward a club of carbon markets. *Climatic Change*. doi:10.1007/s10584-015-1506-z.

Kinley, R. (2016). Climate change after Paris: From turning point to transformation. *Climate Policy*. doi:10.1080/14693062.2016.1191009.

Koester, R. J., & Hall, S. (2014). Leveraging the voluntary carbon market to invest in deeper carbon reduction. *Sustainability: The Journal of Record, 7*(5), 242–246.

Lederer, M. (2011). From CDM to REDD+ What do we know for setting up effective and legitimate carbon governance? *Ecological Economics, 70*, 1900–1907.

Lederer, M. (2012). Market making via regulation: The role of the state in carbon markets. *Regulation and Governance, 6*, 524–544.

Liang, T., Taschini, L., & Palmer, C. (2016). Understanding the demand for REDD+ credits. *Environmental Conservation, 43*(4), 389–396.

Lim, J.-S., & Kim, Y.-G. (2012). Combining carbon tax and R&D subsidy for climate change mitigation. *Energy Economics, 34*, 496–502.

Lo, A. Y., & Cong, R. (2017). After CDM: Domestic carbon offsetting in China. *Journal of Cleaner Production, 141*(10), 1391–1399.

Lubowski, R. N., & Rose, S. K. (2013). The potential for REDD+: Key economic modeling insights and issues. *Review of Environmental Economics and Policy, 7*(1), 67–90.

Mansell, A. (2016). What's ahead for carbon markets after COP21? International Centre for Trade and Sustainable Development – ICTSD. *BIORES, 10(1)*.

Marcello, T. (2011). Voluntary carbon markets: Successes and shortfalls. In Transparency International (Ed.), *Global corruption report: Climate change*. London: Earthscan.

Marcu, A. (2016a, January). *Carbon market provisions in the Paris Agreement (Article 6)* (CEPS Special Report, p. 128).

Marcu, A. (2016b). Governance of carbon markets under Article 6 of the Paris Agreement. In R. N. Stavins & R. C. Stowe (Eds.), *The Paris Agreement and beyond: International Climate Change Policy Post-2020*. Cambridge, MA: Harvard Project on Climate Agreements.

McDermott, C. L. (2014). REDDuced: From sustainability to legality to units of carbon—The search for common interests in international forest governance. *Environmental Science and Policy, 35*, 12–19.

Mehling, M., & Haites, E. (2009). Mechanisms for linking emissions trading schemes. *Climate Policy, 9*, 169–184.

Michaelowa, A. (2014). Linking the CDM with domestic carbon markets. *Climate Policy, 14*(3), 353–371.

Michaelowa, A., & Hoch, S. (2016). How to transition from the CDM to the sustainable development mechanism under the Paris Agreement. *Carbon Mechanisms Review, 1*, 28–31.

Michaelowa, A., & Jotzo, F. (2005). Transaction costs, institutional rigidities and the size of the clean development mechanism. *Energy Policy, 33*, 511–523.

Newell, R. G., Pizer, W. A., & Rami, D. (2013). Carbon markets 15 years after Kyoto: Lessons learned, new challenges. *Journal of Economic Perspectives, 27*(1), 123–146.

Nordhaus, W. (2015). Climate clubs: Overcoming free-riding in international climate policy. *American Economic Review, 105*(4), 1339–1370.

Passero, M. (2010). The voluntary carbon market: Its contributions and potential legal and policy issues. In D. Freestone & C. Streck (Eds.), *Legal aspects of carbon trading: Kyoto, Copenhagen, and beyond*. Oxford/New York: Oxford University Press.

Pettenella, D., & Brotto, L. (2012). Governance features for successful REDD+ projects organization. *Forest Policy and Economics, 18*, 46–52.

Pizer, W. A. (2016). Government-to-Government carbon trading. In R. N. Stavins & R. C. Stowe (Eds.), *The Paris Agreement and Beyond: International Climate Change Policy Post-2020*. Cambridge, MA: Harvard Project on Climate Agreements

Ranson, M., & Stavins, R. N. (2015). Linkage of greenhouse gas emissions trading systems: Learning from experience. *Climate Policy*. doi:10.1080/14693062.2014.997658.

RCCs. (2016, July 31). *Regional non-State actor dialogue on Article 6 of the Paris Agreement*. Regional Collaboration Centres (RCCs) Synthesis Report.

Redmond, L., & Convery, F. (2015). The global carbon market-mechanism landscape: Pre and post 2020 perspectives. *Climate Policy, 15*(5), 647–669.

Rehdanz, K., & Tol, R. S. J. (2005). Unilateral regulation of bilateral trade in greenhouse gas emission permits. *Ecological Economics, 54*, 397–416.

Rezai, A., Taylor, L., & Mechnler, R. (2013). Ecological macroeconomics: An application to climate change. *Ecological Economics, 85*, 69–76.

Sabel, C. F., & Victor, D. G. (2015). Governing global problems under uncertainty: Making bottom-up climate policy work. *Climatic Change*. doi:10.1007/s10584-015-1507-y.

Sandbag. (2016, July). *Bend it, don't break it – Introducing new flexibilities into the EU Effort Sharing Decision*. Sandbag Report.

Schneider, L., & La Hoz Theuer, S. (2016). Design options for reliable and thorough Paris market mechanisms. *Carbon Mechanisms Review, 1*, 19–24.

Schneider, L., Kollmuss, A., & La Hoz Theuer, S. (2016). Ensuring the environmental integrity of market mechanisms under the Paris Agreement. *Stockholm Environment Institute – SEI Policy Brief, 2016.*

Sharma, A. (2016, January 21). *Carbon markets firmly back on the agenda.* International Institute for Sustainable Development – IISD Commentary.

Sorrell, S. (2010). Energy, economic growth and environmental sustainability: Five propositions. *Sustainability, 2*, 1784–1809.

Stavins, R. N. (2016). Market mechanisms in the Paris climate agreement: International linkage under article 6.2. In R. N. Stavins & R. C. Stowe (Eds.), *The Paris Agreement and beyond: International Climate Change Policy Post-2020.* Cambridge, MA: Harvard Project on Climate Agreements.

Sterk, W., Braun, M., Haug, C., Korytarova, K., & Scholten, A. (2006). *Ready to link up? Implications for design differences for linking domestic emissions trading schemes: Joint emissions trading as a socio-ecological transformation* (JET-SET project working paper, I/06).

Stua, M. (2013). Evidence of the clean development mechanism impact on the Chinese electric power system's low-carbon transition. *Energy Policy, 62*, 1309–1319.

Stua, M. (2014, September). *A new paradigm: Establishing a global carbon market as an element for the foundation of a 'Low Carbon Bretton Woods' system.* SPRU working paper.

Swartz, J. (2016, March). *China's national emissions trading system: Implications for carbon markets and trade* (Global Economic Policy and Institutions, Issue Paper 6). ICTSD Series on Climate Change Architecture.

Trotignon, R. (2012). Combining cap-and-trade with offsets: Lessons from the EU-ETS. *Climate Policy, 12*, 273–287.

Tuerk, A., Sterk, W., Haites, E., Mehling, M., Flachsland, C., Kimura, H., Betz, R., & Jotzo, F. (2009a, May). *Linking emissions trading schemes.* Climate strategies synthesis report.

Tuerk, A., Mehling, M., Flachsland, C., & Sterk, W. (2009b). Linking carbon markets, concepts, case studies and pathways. *Climate Policy, 9*, 341–357.

Tulkens, H. (2016, May). *COP 21 and economic theory: Taking stock* (CESifo Working Paper, 5918).

Van der Bergh, J. C. J. M. (2016). Rebound policy in the Paris Agreement: Instrument comparison and climate-club revenue offsets. *Climate Policy*. doi:10.1080/14693062.2016.1169499.

Weischer, L., Morgan, J., & Patel, M. (2012). Climate clubs: Can small groups of countries make a big difference in addressing climate change? *RECIEL, 21*(3), 177–192.

Weischer, L., Warland, L., Eckstein, D., Hoch, S., Michaelowa, A., Koehler, M., & Wehner, S. (2016, May). *Investing in ambition: Analysis of the financial aspects in (Intended) nationally determined contributions*. Germanwatch and Perspectives Briefing Paper.

Well, M., & Carrapatoso, A. (2016). REDD+ finance: Policy making in the context of fragmented institutions. *Climate Policy.* doi: dx.doi.org/10.1080/14693062.2016.1202096.

World Bank. (2016, October). *State and trends of carbon pricing*. World Bank Group Report, Washington, DC.

Zwick, S. (2016, January 02). 6 weeks after Paris, countries moving to build carbon markets. *The Huffington Post.*

Chapter 8
A Hybrid Model to Govern the Mitigation Alliance

Abstract This Chapter provides detail on the Mitigation Alliance's (MA) multi-level, hybrid governance model encompassing demand, supply, and exchange of mitigation outcomes, as well as any other significant aspect of the MA's lifecycle. To begin with, it is necessary to understand how the MA governance model relies upon its unique links with international climate institutions. Subsequently, the analysis is extended to relationships between the MA and non-climate external stakeholders. This stage of analysis involves an explanation of the position and possible impact of the MA within the global context. However, it is also necessary to clarify the MA's internal governance model. Internally, the MA consists of overarching authorities mandated to harmonise the overall organisational structure These authorities guarantee an effective, equitable and transparent functioning of the MA's pillars (the demand, supply, and exchange of mitigation outcomes), in line with the principles and criteria of the Paris Agreement (PA). Once the overarching MA structure has been spelled out, it is then necessary to focus on the specific governance dynamics of each of the pillars. The end result is a hybrid, holistic, harmonised governance model for the MA. The discussion of the MA governance in this Chapter completes the architecture of the MA, thus paving the way for the final Chapters to demonstrate its potential.

Keywords Hybrid governance • Plurilaterally Determined Contributions (PDC) • Members Determined Contributions (MDC) • Governing authorities • Mitigation Alliance (MA) members • Stakeholders

8.1 Introduction

To understand the architecture of the Mitigation Alliance (MA) it is necessary to set out the design of its functioning as well as its governance model. While the functioning of the MA is described in Chaps. 5, 6 and 7 of this book, this chapter is entirely focused on the MA's governance model. The MA governance represents a multi-level hybrid model, where top-down elements mix with bottom-up ones and where a multitude of different actors, or stakeholders, interact at any phase and at any level of the MA functioning. Following this multi-level, hybrid logic is therefore required fully analyse the complexity of the MA governance model.

M. Stua, *From the Paris Agreement to a Low-Carbon Bretton Woods*,
DOI 10.1007/978-3-319-54699-5_8

To begin with it is important to analyse interactions between the MA and some key international treaties and organisations (Sect. 8.2). These include, *inter alia*, the United Framework Convention on Climate Change (UNFCCC) and the Paris Agreement (PA). While both the UNFCCC and the PA represent the foundations of the MA, the PA, thanks to the interpretation of its Article 6 (A6PA) developed in Chap. 4 of this book, also provides its operational guidance. Therefore, while discussing key features of the interactions between the MA and the UNFCCC, Sect. 8.2 once more emphasises the significance of A6PA as the essential part of the accord necessary for any form of experimental climate governance under the UNFCCC. Section 8.2 provides also the rationale for the MA to be implemented in parallel to the UNFCCC. Section 8.3 extends the analysis to other international organisations that may influence the implementation and functioning of the MA discussed in previous chapters. Special attention is paid to the World Trade Organisation (WTO) and to the risks of conflicts between its jurisdiction and that of the now established international climate change regime.

Section 8.4 introduces the proper governance model of the MA. It firstly describes inclusion criteria for MA membership, thus providing a preliminary insight into its composition. This section then briefly describes the various categories of stakeholders that might potentially interact under the MA. While drafting the very first pillars for its stakeholders' interactions, the section provides evidence of the magnitude and complexity of the overall MA architecture and outlines the hybrid model of governance that characterises the MA proposal.

Sections 8.5, 8.6, 8.7, and 8.8 analyse the structure and functioning of the MA authorities both as a whole and separately, describing their governance role for: (a) the distribution of the Net Zero Carbon (NZC) target; (b) the certification of mitigation outcomes; and (c) the outcomes' exchanges within the MA. These sections take a similar perspective to that defined by Levi (1991), emphasising the similarities between the MA authorities and the institutions established by the Bretton Woods.[1]

While the links between the MA and Bretton Woods are further analysed in Chap. 9 of this book, Sects. 8.9, 8.10, and 8.11 describe the governance model of the three pillars which make up the MA in this proposal: (a) the distribution of NZC target among MA members, which creates a system of demand for mitigation outcomes[2]; (b) the certification of mitigation outcomes within the MA, leading to the establishment of a supply side to meet the corresponding demand for mitigation outcomes[3]; (c) the demand/supply dynamics characterising the MA.[4] These sections discuss the interactions of various stakeholders involved in the three pillars, offering

[1] Bretton Woods represented a monetary system aimed to establish common rules for commercial and financial relations among western countries after the Second World War. Operational from 1945 Bretton Woods led to the establishment of the International Monetary Fund (IMF), the World Bank and the World Trade Organization (WTO). The WTO became operational only in 1995. See also Chap. 9 of this book.

[2] See Chap. 5 of this book.

[3] See Chap. 6 of this book.

[4] See Chap. 7 of this book.

an insight into their roles and opportunities within the MA, and further extending the concept of multi-level, hybrid governance as conceived by the MA proposal in this book. As summarised in the concluding Sect. 8.12 of this chapter, this analysis leads to a consideration of the potential effects that may be triggered by the MA, which are fully described in Chaps. 9, 10, and 11 of this book.

8.2 Interactions Between the Mitigation Alliance and the International Climate Regime

As already emphasised in Chap. 4 of this book, the legal provisions included in A6PA provide the necessary regulatory framework for the MA. The MA therefore stands in close relationship with the PA and, more generally, with the UNFCCC. At the same time, due to features determining the institution of experimental governance models within A6PA, the MA maintains an autonomous structure, which runs parallel to most traditional UNFCCC bodies and bonds. A6PA therefore combines *top-down* ties linking the MA to the UNFCCC with *bottom-up* relationships to international climate-related institutions, (as already envisaged, in the case of the latter, by the climate club literature).[5] In this way, A6PA supports the establishment of a multi-level, hybrid governance model, perceived by the most recent climate policy literature as the ideal architecture to guarantee equity and effectiveness for efforts aimed at combating climate change at any level (Weischer et al. 2012; Green et al. 2014; Andresen 2015; Michaelowa 2015; Bodansky et al. 2016). To emphasize the significance of A6PA as a bridge linking the MA and the UNFCCC requires a recap of the key findings of Chap. 4 of this book, in light of additional information and discussions outlined in Chaps. 5, 6, and 7.

A6PA not only includes specific criteria for the implementation of cooperation activities under its framework, but also (directly and indirectly) refers to several other parts of the PA, therefore implicitly requiring its cooperation activities to be aligned with the PA's general principles and objectives. More specifically, the key criteria explicitly included in A6PA can be summarised in the following five points, when applied to the MA:

1. MA members adopt it to implement their Nationally Determined Contributions (NDCs) or those of their reference parties, to allow for higher ambition in their mitigation and adaptation actions and to promote sustainable development and environmental integrity (Paragraph 1 of A6PA). This first criterion leads to the institution of the MA's PDC.
2. MA members shall, while implementing their collaborative actions, ensure environmental integrity and transparency, including in governance, and shall apply robust accounting to ensure, *inter alia*, the avoidance of double counting

[5] See Sect. 3.6 of Chap. 3 of this book.

(Paragraph 2 of A6PA). This second criterion serves as foundation for the system of distribution and enforcement of the MA target among its members.[6]

3. The MA shall be able to establish a mechanism that: (a) promotes the mitigation of carbon[7] emission while fostering sustainable development; (b) incentivizes and facilitates the participation of public and private sectors in carbon emission's mitigation (or simply mitigation) efforts; (c) contributes to mitigation for members hosting the mitigation activities, or to fulfil the mitigation objectives of members acquiring the corresponding outcomes; (d) to deliver an overall mitigation in global carbon emission (Paragraph 4 of A6PA). This third criterion serves to establish the Mitigation Alliance Mechanism (MAM).[8]
4. MA members hosting mitigation activities shall not use the corresponding outcomes for their NDCs when other members acquire their title with the same aim (Paragraph 5 of A6PA). This fourth criterion identifies the use of the MAM as source of certification to track the mitigation outcomes and establish their ownership, therefore justifying the establishment of the Certified Mitigation Outcomes (CMO).[9]
5. MA members shall also adopt non-market approaches[10] aimed at: (a) promoting mitigation and adaptation ambition; (b) enhancing public and private sector participation in the implementation of nationally determined contributions; (c) enabling opportunities for coordination across instruments and relevant institutional arrangements (Paragraph 8 of A6PA). This last criterion supports the recognition of multiple approaches for CMOs exchanges within the MA, including non-market ones.[11]

As a consequence of its special relationship with A6PA, the MA shall follow five key principles embedded in the PA and promoted by the UNFCCC, including: (a) higher ambition in mitigation, leading to an overall increase in global mitigation ambitions; (b) promotion of sustainable development and environmental integrity; (c) maintenance of high standards of transparency and avoidance of double counting; (d) promotion of public and private sectors' engagement in the activities triggered by the MA; (e) effective support for non-market approaches.

These five principles represent the foundation of the links between the MA on one side and the UNFCCC and PA on the other. The connection is further maintained and enhanced by an additional element suggested in A6PA: the authority and guidance of the Conference of the Parties serving as the meeting of the Parties to the

---

[6] See Chap. 5 of this book.

[7] Carbon Dioxide (CO_2) is the most common Greenhouse Gas (GHG) and is often used as a reference unit to express levels of GHG emissions (expressed in CO_2 equivalent, or CO_2eq). In the context of this Chapter and, more in general, of this book the term carbon, expressing CO_2, and the term GHG are synonyms, hence carbon is used as single term to express GHG.

[8] See Chap. 6 of this book.

[9] See Chaps. 6 and 7 of this book.

[10] In the context of this chapter, and more in general of this book, the term 'approach' refers to any form of mitigation exchange.

[11] See Chap. 7 of this book.

Paris Agreement (CMA) over the establishment of the MAM. While this aspect does not automatically translate into a direct competence of the CMA in governing the MAM functioning, it is established by Paragraph 7 of A6PA that the CMA authority shall define the rules, modalities and procedures of an apparatus such as the MAM. As a consequence, the MA cannot autonomously determine the minimum standards concerning its CMO, for these require international recognition under the UNFCCC and PA umbrella.

Meeting the criteria and principles resulting from the links between the MA, the UNFCCC, and the PA requires that the MA: (a) guarantee adherence with the PA's objectives, with special reference to the concepts of ambition and differentiation in a regime of Common But Differentiated Responsibilities and Respective Capabilities (CBDR-RC) applicable to all (Garibaldi 2014; Winkler and Rajamani 2014; Rajamani 2016); (b) safeguard transparency as envisaged by Article 13 of the PA (Hare et al. 2010; Bodansky 2016; Rajamani 2016); (c) implement Monitoring, Reporting and Verification (MRV) systems in line with the standards adopted by the PA and the UNFCCC (Bellassen and Cochran 2015; Chang and Bellassen 2015).

As shown in the previous Chapters of this book, the MA satisfies all three above-listed requirements, therefore meeting the criteria and principles established through in connection with the PA and the UNFCCC. In detail, the MA guarantees adherence to the PA ambition by adopting a long-term NZC target, necessary to align to the global average temperature increase no higher than 2/1.5 °C decreed by Article 2 of the PA. While maintaining its effectiveness through the NZC target adoption, the MA grants equity in distributing the target through the threshold and flexible formula introduced in Chap. 5 of this book. In addition, the MA satisfies the transparency requirements of Article 13 of the PA by adopting members' registries of carbon emission, in accordance with the standards defined by the UNFCCC.[12] More specific CMOs registries shall further enhance transparency MRV process implementation within the MA, hence guaranteeing double counting avoidance, as required by Paragraph 2 of Article Paragraph 2 of A6PA. Finally, by following the UNFCCC-determined standards concerning CMOs issuance and MRV as required by Paragraph 7 of A6PA, the MA guarantees the accountability of its MAM and of the corresponding CMOs.

The strong ties linking the MA to the overall UNFCCC architecture, serve, above all, to guarantee the effectiveness of the MA as a tool to promote and implement the ambitious objectives of the PA. Secondly, the links help consolidate the MA as a single, harmonious structure, providing it with well-defined principles and criteria to avoid any form of fragmentation. Finally, they provide the MA with international recognition, therefore granting legitimacy and reliability to its implementation. Such elements are a recurring feature in the climate clubs literature,[13] where the need for strong ties between clubs and an international climate regime finds its

[12] Serving as public references for the status and progress in MA members' mitigation activities, the registries would offer a significant contribution to the global stocktake as required by Article 14 of the PA (Raman 2016).

[13] See Chap. 3 of this book.

rationale in the avoidance of fragmentation (Falkner 2016) and in the legitimacy/reliability guarantee (Stewart et al. 2013; Das 2015; Potoski 2015).

By connecting the MA to the UNFCCC and the PA, A6PA simultaneously guarantees the highest degree of autonomy for the MA. The key for determining the MA's autonomy from the international climate regime represented by the UNFCCC is given by Paragraph 1 of A6PA, where it is affirmed that: 'Parties recognize that some Parties choose to pursue voluntary cooperation in the implementation of their nationally determined contributions to allow for higher ambition in their mitigation and adaptation actions and to promote sustainable development and environmental integrity'.[14] While *voluntary cooperation* represents by far the most important founding element of any experimental climate governance implemented through A6PA, it is the concept of *some parties* that dignifies the autonomy of such governance from the UNFCCC system. Required by parties that expressed no intention to develop collaborative activities for mitigation (Marcu 2016), the discrimination between parties wishing to voluntary cooperate for higher ambition and those parties that do not implies *de facto* that any *voluntary cooperation* be implemented in parallel to the UNFCCC. The multilateral, international nature of the UNFCCC regime implies that its decisions and actions be addressed with no discrimination to any signatory member. As a consequence, any form of *voluntary cooperation* that only applies to *some parties* cannot be managed and operationalized by the UN system, thus requiring an autonomous structure to take charge of such operationalization.

A6PA provides additional guidance regarding an autonomous governance model applicable to parties (and possibly other authorities) that freely decide to cooperate for higher ambition running in parallel to the UNFCCC. For instance, when referring to actions and strategies implemented through *voluntary cooperation*, Paragraph 2 of A6PA states that these actions must be 'consistent with guidance adopted by the CMA'.[15] The terminology *consistent with* provides legitimacy to the idea of a structure that can act in parallel to the UNFCCC system, as long as its consistency with the UNFCCC is maintained and guaranteed. In addition, A6PA grants to the participating Parties full discretion as to the use of mitigation outcomes produced by voluntary cooperation,[16] therefore emphasising their autonomous role when acting under its framework. Finally, when stating that the mechanism established under the authority of the CMA 'shall be supervised by a body designated by the CMA',[17] the A6PA opens a significant window of opportunity for this body to be external to the UNFCCC, albeit under its supervision.

The high degree of autonomy provided to any voluntary cooperation established under the framework of A6PA also entails significant flexibility for the institution and implementation of such cooperation. Previous Chapters of this book have already emphasised some of the key aspects to such flexibility when applied to the

[14] Paragraph 1 of A6PA.

[15] Paragraph 2 of A6PA.

[16] Paragraph 3 of A6PA.

[17] Paragraph 4 of A6PA.

MA proposal. Firstly, this flexibility provides members that voluntarily decide to join the MA the opportunity to define a common NZC target, a common pathway for distributing the target,[18] as well as a common, single system for the certification and issuance of mitigation outcomes required by members to demonstrate their effective compliance with the established target and corresponding assigned shares of mitigation burden or quotas.[19] Secondly, such flexibility provides the opportunity to extend MA membership to authorities other than traditional parties.[20] Thirdly, it stimulates wide participation of a multitude of stakeholders at various stages of MA target implementation.[21] The MA proposal reflects the consequences of such high flexibility and autonomy in its architecture.

In conclusion to this section, it is pertinent to highlight how the establishment of MA may impact upon and and possibly solve some of the lock-ins that have traditionally limited the efficacy of international action against climate change under the UNFCCC. The innovation produced by the MA can be summarised in two main points:

1. By representing a voluntary system for cooperation, the MA automatically excludes parties that do not wish to do so. Exclusion also applies to any decision-making process within the MA, thereby overcoming constraints of international law that have repeatedly affected the negotiation and implementation processes within the climate change policy framework (Streck 2012). In detail, the voluntary system of cooperation excludes the option of vetoing by actors who aim for low ambitions and low-profile compromises within the UNFCCC (Madden 2014; Caparrós 2016; Clémençon 2016). This element, coupled with the common and defined NZC target accepted by MA members, as well as an equally common and defined system to assign the corresponding share of mitigation burden or quota, allows for the MA to effectively aim for qualities outlined in Chap. 3 (speed, ambition, participation and equity) and identified by climate policy literature as the critical elements for a successful transformational governance in climate change (Biermann et al. 2009; Weischer et al. 2012; Garibaldi 2014; Bodansky et al. 2016; Falkner 2016; Hermwille 2016).
2. By adopting a common target and corresponding distribution system, together with the establishment of a single, shared mechanism to certify and issue mitigation outcomes required to meet the target, the MA reframes the pledge-based NDC implementation system described by Article 4 of the PA effectively aligning the system with the PA's overall objectives. For instance, the MA can lead to the establishment of a PDC regime, with the respective members' NDCs converging into a single framework characterised by a commonly defined target, distribution system and certification mechanism, while still maintaining full

[18] See Chap. 5 of this book.

[19] Quota is used in this Chapter and more in general in this book as synonym of 'assigned share of mitigation burden'.

[20] See Chap. 4 of this book.

[21] See Chap.6 and 7 of this book.

independence in determining their strategies and policies for the full implementation of the contribution.[22] Such a pledge-based implementation system reframing can effectively contribute to solving limitations resulting from lack of ambition and coordination, which have already affected the NDC-based system since its very early stages (Buhr et al. 2014; Geden 2015; Peters et al. 2015; Sabel and Victor 2015; Clemençon 2016; du Pont et al. 2016; Morgan 2016; Rogelj et al. 2016).

As a result of the combination of its ties to the UNFCCC and its high degree of autonomy from the same authority, the MA exhibits a hybrid relationship towards the international policy regime on climate change and a 'win-win' combination from various points of view. On the one hand, the ties guarantee for the MA to operate in a homogeneous, standardised and internationally recognised framework. This avoids fragmentation and provides legitimacy for its aims and implementing tools. On the other hand, the MA's autonomy increases its capacity in terms of ambition, differentiation and coordination, thereby significantly enhancing its overall efficacy and effectiveness.

8.3 Interactions Between the Mitigation Alliance and Other International Institutions

2016 represented a milestone for a plethora of international negotiations whose most recent developments have been significantly affected by the PA. For two of them, the Montreal Protocol on the ozone layer and the mitigation agreement reached by the International Civil Aviation Organisation (ICAO), it is worth analysing possible links and overlaps with the MA proposal. Both cases represent an interesting evolution of the nuanced interpretation provided by the PA to the principle of CBDR-RC. The Montreal Protocol represents one of the highest achievements in terms of international environmental negotiations. It led to a global agreement to phase-out chemicals affecting the ozone layer. Yet, the induced effects of the phasing-out process corresponded to a significant increase of chemical by-products, such as Hydrofluorocarbons (HFC), which represent a highly impacting category of GHG.[23] The long-lasting debate on the limitations and contradictions of keeping HFC under the UNFCCC umbrella (Wara 2008; Böhm et al. 2015) finally ended with the amendment of the Montreal Protocol concerning HFC abatement in October 2016 (Johnston et al. 2016). This historical deal established that the HFC phasing-out processes fall under the Montreal Protocol jurisdiction. Excluding the HFC debate from the discussions surrounding the PA greatly facilitates the implementation of the Montreal Protocol. At the same time, it has only been made possible by the novelties included in the PA. For instance, the progress from an

[22] See also Chaps. 5, 6 and 7 of this book.

[23] See also Sect. 6.5 of Chap. 6 of this book.

ideological discrimination between blocks of countries 'in charge' of acting and countries with no duty to act, which characterised the Kyoto Protocol (KP) through the differentiation between Annex I and non-Annex I parties,[24] towards the new PA regime of responsibilities 'applicable to all', facilitated the evolution of other international negotiation tables in directions aligned with the PA's principles. As a consequence of these most recent evolutions within the Montreal Protocol, it is likely that HFC-related emissions will be excluded from the scope of the PA in general, as well as evolutions such as the MA.

In a similar manner to those of the Montreal Protocol, negotiations for a common ICAO climate change strategy significantly progressed thanks to the novelty of the nuanced interpretation to CBDR-RC provided by the PA. Irreconcilable positions regarding the mitigation burden distribution between ICAO members had for several years hampered any positive outcome, with proposals of duty differentiations among carriers of different nationalities being vetoed one after another. The new CBDR-RC interpretation provided by the PA offered the opportunity for a radical change of perspective in the ICAO negotiations, shifting the differentiation of responsibilities from the carriers to the areas of destination/origin of the corresponding flights.

As a consequence, a common distribution of responsibilities has been finally accepted and all the carriers composing ICAO enjoy equal treatment under the Carbon Offsetting and Reduction Scheme for International Aviation (CORSIA) approved by the International Civil Aviation Organization on October 2016 (ICAO 2016).[25] CORSIA, compared to the Montreal Protocol, is much more likely to interact with, or at least be affected by, the implementation of the MA. While the amended Montreal Protocol concerns a type of GHG that is now likely to be excluded from the basket of gases targeted by the MA, CORSIA refers to the same emissions as the MA. This implies that CORSIA is likely to be implemented alongside the MA, as its founding criteria are not compatible with the MA's. The two systems may become *de facto* competitors for the attraction of mitigation outcomes. If the MA proposal is implemented effectively it is possible to envisage mitigation demand within the MA exceeding demand related to the CORSIA arrangements. As a result, dynamics of CMOs exchange are likely to influence the prices of mitigation outcomes exchanged within the CORSIA regime.

Of all the relations between the MA and international institutions, that between the MA and the UNFCCC represents by far the most relevant and necessary to analyse. However, the relationship of the MA to the World Trade Organisation (WTO) and its related General Agreement on Tariffs and Trade (GATT) is also significant and worthy of careful consideration. Risks and uncertainties concerning climate-trade interactions have characterised the international climate change debate since its earliest beginnings (Doelle 2004; Knox 2004). These have contributed to a variety of disputes between jurisdictions aimed at supporting climate-oriented policies

---

[24] See Chap. 2 of this book.

[25] Exceptions are included for carriers flying from/to some special categories of countries, such as least developed countries and land-lock countries, in accordance to the CBDR-RC principle.

and the overarching architecture of the global trade regime. More specifically, contentious issues concern the application of Border Carbon Adjustments (BCA), the adoption of mitigation-supportive subsidies, and the removal of trade barriers for climate-friendly goods and services (Zelli and van Asselt 2010; Droege 2011).

Establishing BCA, in the form of a penalty applied to products imported from outside a regime governed by specific carbon emission regulations, represents by far the most contentious issue between the newly established climate architecture (MA) and the norms governing the GATT/WTO. Considered the ideal 'stick' to stimulate wider participation in club-style initiatives (Helm et al. 2012; Nordhaus 2015; Victor 2015), the BCA may *de facto* be identified as protectionism conflicting with GATT/WTO principles. Several academics and lawmakers nevertheless identify in GATT Article XX on General Exceptions the windows of opportunity for the implementation of BCA policies in support of mitigation strategies developed within club structures such as the MA (Zelli and van Asselt 2010; Das 2015). More specifically, windows may potentially open through Paragraphs (b), concerning measures necessary to protect human, animal or plant life or health, and (g), referring to measures relating to the conservation of exhaustible natural resources.[26]

While BCA experiments have yet to be implemented within the UNFCCC, similar views are contrasted with the different interpretations that can be given to the 'chapeau'[27] to Article XX, which states that trade sanctions based upon Paragraphs (b) and (g) of Article XX measures 'are not applied in a manner which would constitute arbitrary and unjustifiable discrimination between countries where the same conditions prevail' (Maruyama 2011; Zhao 2013; Mavroidis and de Melo 2015).

Including emissions embedded in goods imported from non-member areas in MA members' emissions data, described in Chap. 5 of this book, allows the MA to set up a non-monetary but effective BCA to avoid carbon leakage.[28] Other interesting elements contained in the MA proposal may help prevent disputes at GATT/WTO level. With no claim to be exhaustive on a subject that would require the concrete design of new trade measures, which certainly falls beyond the scope of this book, this MA proposal still suggests a possible solution to some of the most consistent conflicts between the GATT/WTO jurisdiction and the new international climate policy regime.

More specifically, following the interpretation of GATT Article XX(g) given by Weischer et al. (2012), the MA firstly identifies in, formulated as its aggregate

[26] It is relevant to point out that for the GATT such measures can be effective only in conjunction with restrictions on domestic production or consumption, represented by the NZC target and its distribution among MA members.

[27] Within the WTO legal phrasing 'chapeau' must be understood as the introductory clause that emphasises the manner in which the measure in question has to be applied.

[28] Including this kind of emissions in the aggregate levels of MA members' emissions would imply no advantage for stakeholders outsourcing their industries to profit from carbon leakage, as re-imports would eventually be counted in the aggregate emissions levels of importing members, therefore still impacting the members in the distribution of the NZC target (see also Chap. 5 of this book).

members' carbon budget,[29] an exhaustible natural resource to which specific restrictions on domestic production and/or consumption apply within the MA.[30] Secondly, the MA envisages the institution of a system to protect this exhaustible resource based upon the mechanism introduced by A6PA, which only applies within its jurisdiction. Thirdly, thanks to its inclusion principles, based upon the concept of voluntary cooperation described in Paragraph 1 of A6PA, the MA does not oppose *a priori* discrimination in applying its system, whilst attempting to attract non-participating entities.[31]

By following these three steps, the MA establishes a specific framework where, in principle, climate-related actions and strategies developed within its jurisdiction do not collide with the GATT/WTO international legislation. Their practical application involves, firstly, the establishment of an aggregate peak year for MA, for instance the aggregate floor level to MA carbon emissions represented by the *MAb* term introduced in Sect. 5.2 of Chap. 5 of this book, which defines and quantifies the exhaustible resource to carry atmospheric carbon capacity in line with the 2/1.5 °C target; secondly, the design of an aggregate pathway aimed at safeguarding the exhaustible resource and their translation into quotas[32]; and thirdly, the MA provides its members with a single system to meet their quotas through the adoption of the MAM and its CMO.[33] Following this process, the exhaustible resource becomes a quantified and limited commodity whose exploitation shall be controlled by the MA through certified mitigation actions. By establishing the exhaustible resource and the corresponding preservation system the MA triggers GATT Article XX(g), aimed at the conservation of exhaustible natural resources. Adopting specific restrictions on its production and/or consumption under the MA jurisdiction satisfies the second criteria of Article XX(g) (see note 25). Finally, it is recognised that for any non-member the possibility still exists to enter the MA at any time. Integration would occur by way of non-member formal acceptance of the principles and criteria governing the MA recognition for carbon as exhaustible natural resource, therefore overcoming any possible objection originated by the 'chapeau' to GATT Article XX.

This interpretation of carbon creates the legal foundation not only for the establishment of BCA to safeguard the exhaustible resource/commodity, but, most importantly, for the implementation of a 'common carbon market',[34] whose norms and regulations apply only to MA members.[35] The lifecycle of CMO represents the key element of this 'common carbon market'. Produced only within the MA jurisdiction and certifying MA members' mitigation, CMOs remain within the border of

[29] Relating to atmospheric carrying capacity in line with the 2/1.5 °C target,

[30] With the restrictions corresponding to the NZC target and its distribution as described in Chap. 5 of this book.

[31] See Chap. 5 of this book.

[32] See Chap. 5 of this book.

[33] See Chap. 6 of this book.

[34] Referred as the Mitigation Alliance Carbon Market (MACM) in Chap. 7 of this book.

[35] See Sect. 7.3 of Chap. 7 of this book.

the MA for their entire lifecycle (production, exchange, final banking). This exclusive (yet common) carbon market represents the key to guaranteeing excludable benefits for MA members,[36] a prerequisite for MA implementation according to the climate clubs literature (Potoski 2015; Falkner 2016).[37] Non-members are implicitly excluded from accessing the MA market. At the same time, however, non-members shall be entitled to access the market by accepting its instituting principles, by recognising carbon as exhaustible natural resource (in line with atmospheric carrying capacity relating to limiting global average temperature rise to 2/1.5 °C) and by accepting the rules applied to the MA members in terms of NZC target and corresponding quotas.

8.4 The Mitigation Alliance Structure

The hybrid nature of the MA implies for its structure to be determined by multiple stakeholders at various levels. While describing these stakeholders and their possible interactions is of primary relevance to understanding the general structure of the MA. The MA's hybrid model requires the identification of regulations that interact in defining it. Five main regulation sources can be identified (see Table 8.1): (a) the overall UNFCCC/CMA jurisdiction; (b) the A6PA criteria; (c) the regulations commonly defined by MA members; (d) the individual MA members' jurisdictions; (e) the individual stakeholders' jurisdictions (if and when stakeholders hold normative rights over specific domains). While norms delivered by single members and stakeholders apply only to their specific jurisdictions, regulations relating to (a), (b) and (c) apply to the entire MA structure.

Table 8.1 shows that the UNFCCC, the CMA, and A6PA clearly provide general criteria and standards for the MA structure by defining its scope and rationale. MA members provide most legislation regulating the MA. In the MA proposal, members collectively decide on a variety of issues and norms, which require the institution of a MA General Assembly (MAGA). The MAGA represents the first authority that the MA requires for its own functioning. Individual members and individual stakeholders holding jurisdiction over defined domains can additionally legislate on policy-oriented subjects.

The framework established by this multi-level regulation system directly and indirectly affects a plethora of stakeholders. Focus is here given to stakeholders that are potentially directly affected by, as well as interact within and therefore constitute, this multi-level MA regulation system. Table 8.2 provides an overview of key stakeholders directly involved in the different levels of MA governance and functioning.

---

[36] For the benefits' analysis see Chap. 10 of this book.

[37] See also Chap. 3 of this book.

Table 8.1 MA regulative resources and respective competence fields

UNFCCC, CMA and single parties	A6PA criteria	Regulations commonly defined by MA members	Individual MA members' jurisdictions	Other stakeholders' jurisdictions
General mitigation objectives	Voluntary participation requirement	Acceptance of carbon as natural exhaustible resource	Internal distribution of quotas	Internal distribution of quotas
Transparency requirements	Higher ambition mandate	Definition of the authorities in charge of managing the MA	Exclusion of mitigation sectors and categories	Exclusion of mitigation sectors and categories
Differentiation requirements	Sustainable development mandate	Aggregate peak year establishment	Adoption and/or exclusion of CMOs exchange approaches	Adoption and/or exclusion of CMOs exchange approaches
Emission registry regulation	Double counting avoidance	Aggregate target definition	Integration of MA-related strategies in wider policies	Integration of MA-related strategies in wider policies
MA membership for non-Party entities	Environmental integrity	Target distribution in form of quotas	Additional share of proceeds definition	Additional share of proceeds definition
Guidance for the establishment of the certification mechanism	Establishment of certification mechanism	Norms regulating the enforcement of MA rules		
Minimum CMO standards definition	Stimulating participation by non-Party public and private stakeholders	Design of certification mechanism norms		
Minimum Monitoring, Report and Verification (MRV) standards definition	Guarantee for overall mitigation	Exclusion of mitigation sectors and categories		
Initial share of proceeds definition	Inclusion of non-market approaches	Additional CMO certification standards		
		MRV standards definition		
		Double counting avoidance registries		
		Norms regulating the MA carbon market		
		Design of rules towards non-members		

Table 8.2 Categorisation of stakeholders directly involved in the MA structure and dynamics

	(1) MA institution and participation	(2) MA target – demand	(3) MA certification – supply	(4) MA exchange dynamics
(1) Stakeholders external to the MA	UNFCCC	UNFCCC	UNFCCC	UNFCCC
	CMA	CMA	CMA	CMA
	Non-MA members	Intergovernmental Panel on Climate Change	Non-MA members	Non-MA members
	Universities, research centres, scientific institutes, think tanks	Auditing entities	Other UN agencies	Other UN agencies
		Universities, research centres, scientific institutes, think tanks	International, regional and sectoral organisations	International, regional and sectoral organisations
			Auditing entities	Auditing entities
			Non-governmental organisations	Non-governmental organisations
			Universities, research centres, scientific institutes, think tanks	Universities, research centres, scientific institutes, think tanks
(2) Public authorities internal to the MA	MA members	MA authorities	MA authorities	MA authorities
	MA authorities	MA members	MA members	MA members
	Non-MA members parties	National, regional, local authorities and cities	National, regional, local authorities and cities	National, regional, local authorities and cities
	Universities, schools, research centres, scientific institutes, think tanks	Public companies	Public companies	Public companies
		Public utilities	Central banks	Central banks
		Universities, schools, research centres, scientific institutes, think tanks	Banks and financial institutions	Banks and financial institutions
		Other public authorities	Universities, schools, research centres, scientific institutes, think tanks (including technology developers)	Universities, schools, research centres, scientific institutes, think tanks (including technology developers)
			Other public authorities	Other public authorities

(3) Private stakeholders internal to the MA	MA members	MA members	MA members	MA members
	Universities, schools, research centres, scientific institutes, think tanks	Auditing entities	Banks and other financial institutions	Banks, brokers, financial intermediaries and other financial institutions
		Universities, schools, research centres, scientific institutes, think tanks	Firms, companies, private utilities and sectoral organisations	Firms, companies, private utilities and sectoral organisations
			Auditing entities	Auditing entities
			Law firms	Law firms
			Consultancy agencies	Consultancy agencies
			Universities, schools, research centres, scientific institutes, think tanks (including technology developers)	Universities, schools, research centres, scientific institutes, think tanks (including technology developers)
(4) Non-defined stakeholders internal to the MA		Traditional communities	Traditional communities	Traditional communities
		Non-governmental organisations	Non-governmental organisations	Non-governmental organisations
		Movements and civil society groups	Movements and civil society groups	Movements and civil society groups
		Individuals	Individuals	Individuals

Universities, research centres, scientific institutes and think tanks appear in different rows as well as columns. This is explained in more detail below, along with other stakeholders appearing in different columns. Organised in four categories representing four different levels of governance, Table 8.2 provides an insight into stakeholders potentially involved in MA governance and functioning. While the four levels of action are analysed in greater detail in other sections of this Chapter, this section provides some guidance to help interpret Table 8.2.

With several stakeholders recurring in Table 8.2, the overall architecture of the MA appears like a dynamic and multi-layered network of interactions, where stakeholders' roles can radically change depending on specific interactions. Its composition includes, alongside institutions responsible for the MA, public authorities at different levels, multiple categories of private stakeholders from civil society, reflecting the multi-polarity of climate action in recent years (Gallaraga et al. 2011; Böhmelt 2013; Lund 2013 Downie 2014; Hale 2016; Jacobs 2016). However, distinguishing between centralised and decentralised governance rules, as well as between top-down and bottom-up approaches, seems meaningless in this context. As a consequence, the MA system is characterised by multi-polar governance dynamics where a variety of approaches and sub-systems, which are differently implemented by the stakeholders involved, form a nested governance system (Wyborn and Bixler 2013; Bradshaw 2014; Kashwan and Holahan 2014; Andresen 2015).

Two significant exceptions to the MA's multi-polar governance are represented by the UNFCCC/CMA dichotomy and by the group including universities, schools, research centres, scientific institutes and think tanks. The UNFCCC/CMA appears recurrently as a result of its unifying principles, objectives, criteria, rules and standards, which form the basis for the architecture of the MA. Its overarching presence contrasts the risks of fragmentation, identified in climate policy literature as one of the biggest limits of multi-level climate governance (Biermann et al. 2009; Andresen 2015; Well and Carrapatoso 2016), which would otherwise be formed of a multitude of stakeholders with different, if not contrasting private interests. At the same time, this binomial presence, together with the institution of MA governance authorities, encourages transnational cooperation, which has been envisaged as the ideal pathway for effective climate governance (Andonova et al. 2009).

The group of universities, schools, research centres, scientific institutes and think tanks appears in several cells in Table 8.2 because this group engages in providing scientific evidence as well as in proposing technically and politically feasible solutions to climate change. These stakeholders play a significant role at any stage of MA implementation. Universities, research centres and scientific institutes analyse climate-related issues, increase understanding across life and social sciences, develop solutions and disseminate related findings. Universities, research centres and scientific institutes, including those embedded in private firms, also research and develop technological, social, financial and organisational innovations required to cope with the ambitious goals of the MA and climate change in general. Together with schools, universities also build capacity to provide future generations with tools and skills necessary to facilitate the transformational process envisaged by the

MA proposal.[38] Traditionally focused on scrutinising and advocating research findings, think tanks play a more specific role in the development of the MA proposal. They may be categorised as catalysers and promoters of ideas and, more than any other stakeholder included in the group analysed in this section, they can be subject to and guided by policy pressures.

8.5 The Structure of the MA Governing Authorities

Before analysing stakeholders directly involved in the three pillars of the MA (demand, supply and exchange of mitigation outcomes), as well as the dynamics governing their interactions, it is first necessary to discuss the characteristics of those institutions necessary to guarantee successful MA governance. Due to specific features regarding its goals and functioning, this book suggests that the MA governance model might succeed by adopting a Bretton Woods-style architecture in line with the ideas proposed in a visionary paper by Maurice David Levi (Levi 1991).[39]

This section introduces the structure and competences of the authorities necessary to ensure an effective, equitable and efficient implementation of MA governance. In principle, the MA requires the establishment of a set of interrelated bodies mandated to oversee the correct implementation of the MA's three main pillars (demand, supply and exchange of mitigation outcomes), and to define regulations listed as 'commonly deliberated by MA members' in Table 8.1. Four authorities may be envisaged for MA governance: (a) the MA General Assembly (MAGA) to act as the MA regulator authority; (b) the MA Target Fund (MATF) mandated to govern the demand for mitigation outcomes; (c) the CMO Bank in charge of certification and issuance of mitigation outcomes; and (d) the CMO Exchange Board (CMO-EB) overseeing the implementation of the CMOs exchange approaches adopted by MA members. Due to coordination and harmonisation implicitly required by the MA rationale and aims, the four authorities can be envisaged as autonomous organs operating under a single, unifying body, rather than as separate but interrelated bodies. Their description below thus follows the 'organs' logic. Yet, the subsequent description can in principle be equally applied even if a 'bodies' logic prevails in the MA set up.

The MAGA shall define norms that, commonly shared by MA members, aim to guarantee the general functioning of the MA. Setting up the other three authorities in charge of the overall governance of the MA is included among its functions; therefore the MAGA can be identified as the founding authority of the whole MA. The MAGA is based on the following five premises:

1. Free and voluntary participation/inclusion of any party and/or entity authorised by its parent party (as for Paragraph 1 of A6PA).

[38] See also Chap. 11 of this book.

[39] The rationale for this association is analysed and discussed in Chap. 9 of this book.

2. Equal representation of all members of the MA.
3. Formal acceptance of criteria and principles resulting from the MA's ties to the PA and the UNFCCC.
4. Application of the unanimity principle in decision-making.
5. Legal binding status of decisions affecting all MA members.

Once established, the MAGA is in charge of developing regulations (see Table 8.1) and managing memberships, and guaranteeing the respect of A6PA principles. In line with these principles, the MA cannot be envisaged as a closed structure, providing instead the opportunity for new members to join over time. Any additional MA member shall therefore be accepted by the MAGA and automatically join it. The MAGA is primarily mandated to formally adopt norms and regulations in line with principles and criteria of the UNFCCC, the PA and A6PA in particular, as described in Sect. 8.2 of this Chapter. In doing so, the MAGA guarantee that the MA fully embrace the legal framework provided by these international regimes, thus defining the scope of action for the entire MA. In the context of collaborative action, and supported by the nuanced interpretation given by the PA to the CBDR-RC concept, all MA members shall be equally represented in the MAGA. Supported by the members' voluntary participation to the MA, unanimity as founding principle for decisions taken by the MAGA shall ensure effectiveness of the general collaborative action feature of the MA. To further guarantee the effectiveness of MAGA decisions, all MA members shall accept these principles as binding and ensure their full application within their own jurisdictions.

Initially, MAGA functions shall be limited to the legislative process required to set up the MA. Over time, its functions will diversify and include amendments to the legislative body, leaving the MA's governance to other authorities envisaged by the MA proposal. When implementing the pillars that form the basis of the MA (demand, supply exchange of mitigation outcomes) it is the responsibility of the MAGA to grant the necessary flexibility to the MA's implementation process so as to avoid specific frictions developing between MA members over time. Although MA governance lies mostly in the hands of the other authorities specified above (MATF, CMO Bank, and CMO-EBO), the MAGA may possibly retain a litigations settlement responsibility so as to help minimise friction between MA members.

8.6 The Mitigation Alliance Target Fund

The MATF defines and distributes the MA target. It is therefore responsible for establishing CMOs demand and is structured similarly to the International Monetary Fund (IMF). A Board of Governors, composed of representatives from each MA member, as well as a corresponding Executive Board (MATF-EB), governs the MATF, whose duties can be summarised in the following four points:

1. To oversee the MA target distribution over the defined timetables by applying the formula described in Chap. 5 of this book.

2. To take charge of CMOs' banking aimed at demonstrating MA members' compliance with their quotas.
3. To guarantee double counting avoidance.
4. To enforce the MA mitigation system.

The MATF Board of Governors shall be in charge of the first three competences. Authority over the last shall be given to the MATF-EB. Similar to IMF rules, representatives from each of the MA members shall compose the Board of Governors. The quota distributed among MA members over time resembles the IMF reserve asset's management (insofar as MA members quota represent assets to be banked within the corresponding timetables). To meet their assigned mitigation, MA members are required to collect and bank a quantity of CMOs corresponding to, or exceeding, their quotas. The MATF therefore holds the dual role of distributing quotas among MA members and of collecting their certified mitigation in the form of CMO, which serve for MA members to demonstrate their mitigation achievements. CMOs collection also requires the institution of a registry by the MATF Board of Governors so as to take account of the CMO banking process and of their corresponding holders. This registry shall act as a tool to avoid double counting, hence serving as an instrument to guarantee the environmental integrity of the MA.

As to what concerns the MA target's enforcement, the MAGA shall be responsible for establishing the norms regulating the enforcement process, while the MATF-EB shall hold the mandate for effectively enforcing them. Based on the IMF Executive Board, the MATF-EB shall include representatives from all MA members, with the distribution of the Board's quotas based upon the emission levels of individual MA members, which is subject to change over time.[40] The MATF-EB decisions may be made unanimously or based upon a qualified quota-based majority, according to the decisions taken by the MAGA concerning the structure of the MATF. The MA proposal envisages four, non-exclusionary enforcement actions to ensure compliance with its target: (a) monetary fee; (b) rating; (c) exclusion; (d) self-empowerment through the recognition of the achieved excludable co-benefits. At the present stage of the MA proposal, the envisaged enforcement actions to ensure compliance with the MA target remain embryonic, requiring both additional research to be fully developed and policy-making decisions to be applied. Nevertheless, this section proposes a brief commentary introducing each of them, aimed at clarifying their rationale and significance within the overall MA.

Sanctioning failure to meet obligations within regimes 'applicable to all', such as the UNFCCC, with a monetary fee, is confronted by the Westphalian Dilemma (Nordhaus 2015), which places the fundamental rights to sovereignty and self-determination of nations above any other form of jurisdiction. As a consequence no effective enforcement system to sanction parties that fail to meet their obligation can be effectively implemented in an international regime context, where parties act as sovereign states. The MA can overcome this limitation as its membership can be composed by both sovereign states and other entities that voluntarily agree to

[40] See also Chap. 5 of this book.

cooperate under the common MA framework. By voluntarily accepting to cooperate, MA members implicitly hold a 'limited sovereignty' status within the MA. As a consequence, the MAGA can establish monetary sanctions against MA members that fail to meet their quotas during the established timetables. The MATF-EB can enforce sanctions by requiring monetary compensation from defaulting members. In principle, this sanction shall be calculated according to the MA market price for CMO at the end of each reference timetable. Quotas that were not met in a specific timetable shall be added to the defaulting members' new quotas in the following timetable. The released funds shall be added to those collected through the MA-level application of the share of proceeds, therefore increasing the MA financial capacity for actions and strategies aimed at climate change adaptation and resilience.[41]

A credit rating system to stimulate MA members' compliance with their quotas might imply that CMOs from the jurisdiction of non-complying MA members be discriminated in terms of value compared to other CMOs. This approach would threaten the entire MA, which is based on the exclusivity of CMO as proof of compliance, as well as on their identity as actual mitigation, regardless of their origins. As a consequence, rating systems discriminating according to success in complying cannot be established within the MA. Nevertheless, a 'pseudo-rating' system can be envisaged where quantities of CMOs corresponding to the unmet quotas in a determined timetable shall be automatically added to the defaulting members' quota in the following timetable. Such a system would discourage investment in activities producing CMOs in defaulting jurisdictions, thereby encouraging MA members to meet their quotas to prevent dis-investment. The MATF-EB shall take charge of collecting CMOs produced by the 'pseudo-rating' system to eventually bank them as shares of the previously non-accomplishing members' quotas.

In principle, exclusion from the MA may represent the most effective system to prevent MA members from failing to meet their quotas. Not limited by international law, and relatively simple if compared to the two previously discussed systems, membership exclusion would discourage MA members from free-riding as doing so would cause them to lose access to MA co-benefits.[42] The non-discriminatory nature of such a system, which, if fully applied, would not take into account any specificity related to the circumstances for a MA member's failure in meeting its quota, may represent the biggest disadvantage of the exclusion system. This limitation may be balanced by decisions taken by the MATF-EB, which may apply a case-by-case evaluation process to decide upon the possible exclusion of a defaulting member.

While sharing some elements with the exclusion system, self-empowerment through excludable co-benefits' recognition represents the ideal and most peculiar form of enforcement envisaged by the MA proposal. Based upon the presumption that any MA member shall be able to recognise and test the effectiveness of MA's co-benefits, the self-empowerment system is by nature unlikely to be enforceable during the MA's very early stages. It is indeed more likely for this self-empowerment process to emerge as a significant enforcement tool once the co-benefits linked to

[41] See also Sect. 7.3 of Chap. 7 of this book.

[42] Chapters 9 and 10 of this book provide a detailed analysis of the MA co-benefits.

MA are evident. Hence it is only likely to apply once the entire MA is set up and running, allowing MA members and their relevant stakeholders to gain experience and confidence in its functioning. The four listed enforcement systems can be non-excludable and, due to the different limitations affecting each of them, the MATF-EB may eventually decide to establish a nuanced single, enforcement system, based upon their mix.

8.7 The Certified Mitigation Outcome Bank

The CMO Bank as envisaged by this MA proposal shall refer to another institution modelled at Bretton Woods: the World Bank. Including representatives from all MA members in proportion to their emission levels, the CMO Bank shall be provided with specific mandates corresponding to the key objectives of the MA and summarised in four points:

1. To stimulate the meeting of MA ambitions.
2. To facilitate adaptation and resilience processes within and possibly beyond MA borders.
3. To increase sustainable development among MA members.
4. To guarantee transparency and accountability in the implementation of any of the three above listed points.

To take charge of its mandates, the CMO Bank shall be responsible for certification and issuance of any CMO, therefore being in charge of the MAM functioning. The CMO bank shall also take full responsibility for managing the MA-level share of proceeds[43] and those funds resulting from the application of a monetary sanction towards defaulting members. Similarly to the Clean Development Mechanism (CDM),[44] the CMO Bank shall be supported by private stakeholders (i.e. accounting and rating firms) in the CMO's certification process, while maintaining total authority over CMO's effective issuance. By certifying and issuing CMOs, which *de facto* represent a financial asset holding an effective value,[45] the CMO Bank financially supports any mitigation activity implemented within the MA jurisdiction and in line with the MA's principles and criteria.

By taking charge of collecting and distributing the MA's share of proceeds outlined in Paragraph 6 of A6PA,[46] the CMO Bank holds the mandate to directly finance

[43] See also Sect. 7.3 of Chap. 7 of this book.

[44] In the Kyoto Protocol's CDM, the Executive Body (EB) in charge for its functioning was supported by private stakeholders, denominated Designated Operational Entities (DOEs), in its Monitoring, Reporting and Verification (MRV) process (see also Chap. 6 of this book).

[45] CMO value is linked to their demand established by the NZC target, as widely discussed and demonstrated in Chap. 9 of this book.

[46] Paragraph 6 of A6PA states that: 'the CMA shall ensure that a share of the proceeds from activities under the mechanism referred to in paragraph 4 of this Article is used to cover administrative

and support adaptation and resilience actions. Although Paragraph 6 of A6PA suggests that the funds could be left to developing country parties that are particularly vulnerable to the adverse effects of climate change to meet the costs of adaptation, the CMO Bank may add, with the approval of the CMA, an additional discrimination by distributing these funds' exclusively among MA members.

In principle, criteria for collecting and distributing the share of proceeds' funds shall be defined by the CMO Bank, with the mandate to be defined in accordance to the criteria and principles of the PA and the UNFCCC. In defining these criteria, the CMO Bank shall potentially develop a tool both to support sustainable development and to ensure equitable collection of the share of proceeds. The CMO Bank may in fact establish variable rates in the share of proceeds according to the sustainable development performance of CMO-rewarded actions. By adopting internationally recognised indexes for sustainability performance, such as Driving Force Indicators – DFI (Mannis 2001), the CMO Bank can (positively) discriminate among activities benefitting from certification and price them accordingly through the application of different shares of proceeds.

As certifier of mitigation outcomes recognised within the MA, the CMO Bank shall establish criteria and standards to guarantee an effective, just, equitable, accountable and transparent CMO certification process. Ideally, the CMO Bank will combine high-level standards with low exchange costs to ensure that small-scale, highly sustainable actions are not excluded, as has been the case in the past where excessive bureaucracy often stymied small-scale certification.[47] Combined with a nuanced balance between transparency and accountability requirements, this may ensure that equity act as a guiding principle for actions producing mitigation outcomes within the MA. The CDM's long-term experience, combined with standards' improvements produced by renowned private stakeholders such as the Verified Carbon Standard (VCS) and the Gold Standard,[48] provide the basis for the CMO Bank to learn from past experience.

8.8 The Certified Mitigation Outcome Exchange Board

The CMO-EB oversees the implementation of CMOs exchange approaches adopted by MA members. It shall be based on a WTO-type structure. The Bretton Woods Agreement provided the foundation for a structure in charge of trade settlements between its members, yet the WTO was founded over 50 years later, officially entering into force on January 1995. As already discussed in Sect. 8.3 of this chapter, the WTO is likely to affect the MA in many ways. The variety of CMOs exchange approaches envisaged for the MA is likely to lead to controversies and frictions

expenses as well as to assist developing country Parties that are particularly vulnerable to the adverse effects of climate change to meet the costs of adaptation' (Paragraph 6 of A6PA).

[47] See also Sect. 7.2 of Chap. 7 of this book.

[48] See also Sect. 7.2 of Chap. 7 of this book.

among MA members, as well as between members and entities external to the MA, over time. Hence the institution of the CMO-EB shall maintain a dual role. It shall be responsible for both the settlements of CMOs exchanges' controversies within the MA and for managing possible disputes with entities external to the MA.

By discussing the possibility for the CMO-EB to oversee disputes between MA members and entities external to the MA, this section provides a brief commentary on interactions envisaged between MA governance models and entities listed as external to MA in Table 8.2. Objectives, goals, aims and targets of the MA shall maintain a strict link with both the UNFCCC and the CMA, with the MA's PDC representing their key link. In addition, the MA shall be aligned with the principles and objectives of the PA. These principles include *inter alia* CBDR-RC, higher ambition, transparency and justice. The relationship between the MA and the multiplicity of United Nations agencies that act in the field of climate change and/or in interrelated areas and sectors shall initially be governed at individual MA members' level, before establishing more direct ties between UN agencies and the MA as an aggregate agent. This second level of interaction with the UN, which shall possibly require direct negotiations for the recognition of the MA's agent status, may enable the MA to directly access UN agencies' support and advice for its effective implementation over time. For similar reasons, the MA shall seek direct relationships with international, regional and sectoral organisations.[49]

Some final considerations concerning actors not directly participating in the MA are required to finalise this section. Based on Paragraph 1 of A6PA, no discrimination other than the self-will shall determine participation in the MA. In principle, any party that has ratified the PA, or any entity authorised by its party, shall be entitled to join the MA to become a member at any time. To ensure the smooth functioning of the MA, joining the MA is only likely to be possible at the end of any determined timetable. Newcomers joining the MA shall be required to fully accept its principles, goals and regulations.

Parties that voluntarily decide to maintain full independence from the MA, as well as MA members that voluntarily leave the MA or are excluded due to their failure in meeting their quotas,[50] will have limitations imposed by the MAGA in accordance with the PA principles and the CMA indications. In principle, entities external to the MA shall hold no right to:

(a) Directly access the CMOs exchange system;
(b) Sell any form of mitigation certified within their domains to MA members, if they are to be used for complying with their quotas;
(c) Directly participate as technology providers, manufacturers, financial intermediaries or in any other form to support activities aimed at meeting the MA target;
(d) Access funds released by the MA share of proceeds, if not established otherwise by the MAGA.

[49] With these organisations including inter alia: the WTO, World Bank, the International Monetary Fund (IMF), the Organisation for Economic Co-operation and Development (OECD), the Asian Development Bank, the World Energy Council (WEC), and the ICAO.

[50] When included as enforcement tool (see also Sect. 8.6).

While exceptions to these limitations may be granted by the MAGA, it is important for these limitations, as well as any other identified by the MA members, to be rigorously applied. This ensures that they maintain a dual role in the overall functioning of the MA. On the one hand, these limitations represent the sources of excludable benefits that encourage active participation in the MA and stimulate the NZC target to be met. On the other hand, these limitations encourage external entities to join the MA, therefore increasing its width and efficacy over the time. Exceptions in the fields of scientific R&D, as well as for accounting firms external to the MA, may help streamline the overall MA functioning, therefore requiring special attention by the MAGA. Scientific cooperation may represent a first step in integrating external entities into the MA, while at the same time enhancing the development of new technologies capable of supporting implementation of ambitious mitigation objectives worldwide. The inclusion of external accounting firms finds instead its rationale in greater transparency and accounting quality possibly provided by these firms.

8.9 Governing the Demand for Mitigation Outcomes

Overseen by the MATF under the guidance of the PA's principles and objectives, and based on the application of the dynamic formula for the MA distribution among the MA members introduced in Chap. 5 of this book, demand for mitigation outcomes within the MA requires participation and inclusion of various stakeholders in a multi-level network. Although this multi-level network represents the main object of discussion for this section, it is also necessary to emphasise how combining the PA's principles and objectives with the dynamic formula will affect the equity criteria as they relate to the more nuanced interpretation to CBDR-RC as introduced by the PA.[51] Some may still object that equity is lacking in the MA proposal due to the absence of explicit references to historical responsibilities, which represented a key element of developing countries' diplomacy in defining CBDR-RC (Den Elzen et al. 2005; Müller et al. 2010; Deleuil 2012; Mattoo and Subramanian 2012; Wei et al. 2014).[52] The relevance of historic responsibility has been drastically reduced in the PA's concept of CBDR-RC 'applied to all' (Winkler and Rajamani 2014). Recent studies support this concept by clearly demonstrating how historical differentiation has lost some of its significance as a result of carbon emissions increases in fast developing countries, coupled with a plateauing of carbon emission levels in developed countries (Goldemberg and Guardabassi 2015).

[51] See also Chaps. 2 and 5 of this book.

[52] See also Chap. 2 of this book.

The formula envisaged for distributing the NZC mitigation target among the MA members[53] can be effective in addressing issues with a delicate topic like equity. Its application requires MA members registering high levels in per capita carbon emissions to take the mitigation lead. At least in the first phases of MA implementation these are likely to be developed countries. The formula also virtually exempts least developed country members from any mitigation obligation and provides them with ample space for increasing their emissions, thus facilitating their development process.

By guaranteeing the environmental integrity of the MA thanks to the readjustment provided by the *Vt* term,[54] the MA proposal offers the opportunity for the MA's least developed members to 'outsource' their carbon emissions increases to more developed members, further increasing the overall embedded equity value. While members from rapidly developing countries are likely to be exempt from mitigation obligations at the beginning of MA implementation, the application of the threshold's discrimination discourages excessive carbon emissions increases, as per capita levels of carbon emissions in that are in excess automatically entail the allocation of a quota.[55] The adoption of 'net' per capita emissions levels to distribute the NZC target represents another drive for equity in MA.[56] The net per capita levels guarantee anybody within the MA an equal right for a minimum quota of emissions.

Additionally, the application of the formula to every timetable established by the MAGA and therefore to new data and carbon emissions levels concerning MA members, represents another effective way to account for differentiation in future quotas' allocation. The formula thus favours members that mitigated their carbon emissions in previous timetables and discourages others from excessively increasing their own levels. These aspects of the MA proposal align with Garibaldi's (2014) concept of 'outcome equity' as opposed to 'historical equity' and shift the equity debate towards the concept of 'future responsibility' (Michaelowa and Michaelowa 2015).

While the MATF represents the overarching authority in charge of both MA mitigation demand governance and the correct application of the dynamic formula, MA members represent its primary stakeholders. MA members are fully responsible for meeting their quotas. While entailing many duties, this responsibility also offers them the opportunity to use their discretion in distributing mitigation requirements within their jurisdiction. This entitles each MA member to develop specific policies and strategies according to their distribution preferences. It appears likely that MA members' decisions regarding such distribution shall impact their NDCs, when members correspond to parties that ratified the PA, or it may impact climate strategies more generally when applying to members that are only a fraction of PA's

[53] See Chap. 5 of this book.

[54] See Sect. 5.3 of Chap. 5 of this book.

[55] See Sect. 5.3 of Chap. 5 of this book.

[56] To distribute the NZC target between MA members with a per capita emissions level higher than the average of the MA, the formula takes into account the portion of per capita above the MA average (see also Sect. 5.3 of Chap. 5 of this book).

signatory parties.[57] NDCs of parties acting as MA members and local climate strategies implemented by non-party MA members are *de facto* comparable under the MA in the form of members' climate policies, therefore being here denominated as Members Determined Contributions (MDCs). Aside from deliberating the internal distribution of the quota, members can also define sectors to be involved in CMOs production, as well as the approaches for their exchange, recognised within their jurisdiction. Such powers represent the foundation of the MDCs.

MA members' governance dynamism regarding the demand for mitigation outcomes can be transferred to other, lower levels. Members' public or private stakeholders, charged with quotas, may decide to further distribute them within their domains and to define strategies for the Monitoring, Reporting and Verification (MRV) process, provided that these stakeholders are aligned with the MA member's decisions, overall criteria and principles. Stakeholders operating at these levels may include *inter alia*: public authorities, private sectoral associations, single firms and utilities. Such stakeholders would simultaneously be objects of decisions taken at higher levels as well as subjects acting on their behalf. This dual dynamic can be maintained until the very 'lowest' level, represented by individuals.

As a result, governance of MA mitigation demand appears, like so many other aspects of the MA proposal, hybrid in nature, where vertical hierarchies based upon jurisdictions are potentially counterbalanced by high degrees of freedom for stakeholders operating within these jurisdictions. The multiplicity of approaches adopted by individual MA members and stakeholders appears to be the most likely consequence of this hybrid governance model, leading to a high degree of differentiation in mitigation outcomes' demand, both between and within MA members' jurisdictions.

8.10 Governing the Supply of Mitigation Outcomes

Compared to the demand side, the governance of mitigation outcome supply implies a more limited number of interaction levels between the different stakeholders. Governed by the rules defining the MAM and CMO and overseen by the CMO Bank, this specific part of the MA governance includes two main levels: (a) the definition of limits and exclusions for CMOs production/issuance; and (b) the actual production/issuance of CMOs. The first level encompasses limitations that apply to all MA members' stakeholders and are decided on by the MAGA. It also includes any decision regarding additional limits to be applied under the jurisdiction of individual MA members as well as decisions at stakeholders' level. Similarly to the governance of demand, the definition of limits and exclusions on the supply side allows MA members to develop their own climate strategies and contribute to the implementation of their NDCs/MDCs. The exclusion of specific sectors and/or

[57] Authorities like states, regions, local communities and cities participate in the form of MA members when authorised by their reference parties, as previewed by A6PA.

types of activities from CMO recognition may depend upon different parameters characterising members' policies concerning climate change, but also their policies on several other priority areas, such as sustainable development, technological innovation, energy security, poverty eradication and other socio-environmental priorities.

Once sectors and types of projects that can be awarded with CMOs for their mitigation outcomes have been established, the governance of their supply shifts towards CMOs production/issuance. Table 8.3 summarises the most relevant stakeholders for CMOs production/issuance and organises them into four categories: (a) CMOs producers; (b) stakeholders supporting CMOs production; (c) certifiers, monitors, reporters and verifiers (MRVers); (d) shareholders.

These categories of stakeholders directly involved in governing the supply of CMOs resemble features of the CDM and other -reduction related mechanisms and the Reducing emissions from deforestation and forest degradation (REDD+) regimes. Since its first application, the CDM, which plays a significant reference role for the MAM and related CMOs,[58] has been characterised by the interaction of a multitude of stakeholders that fits the four categories envisaged for the MA (Streck 2004; Haites et al. 2006; de Coninck et al. 2007; van der Gaast et al. 2009; Zavodov 2012; Bayer et al. 2013; Methmann 2013; Gandenberger et al. 2015; Kuchler 2015; Watson et al. 2015). The same also applies to REDD+ (Thomson et al. 2011; Pettenella and Brotto 2012; Shiroyama et al. 2012; Vant and Vedeld 2013; Kashwan and Holahan 2014; Well and Carrapatoso 2016).

However, the MA proposal offers some significant innovations compared to the CDM and REDD+ regimes,[59] involving a greater range of stakeholders in its mitigation outcome supply's governance. Firstly, by merging three traditional types of mitigation production (reduction, avoidance and sequestration) under a single certification mechanism (the MAM), the MA unites stakeholders previously operating under different regimes into one framework. Secondly, with the MAM offering a single offsetting system that includes any form of CMOs exchange, the MA guarantees all stakeholders involved in the supply of mitigation outcomes to be potentially recognised as such in every members' jurisdiction. As a consequence, the CMOs supply's governance is more clearly defined compared to similar regimes in the past.

The category of CMOs producers potentially includes both private and public authorities that can directly produce mitigation outcomes such as *inter alia*: public and private manufacturers from any industrial sector; energy utilities and firms; transport-related stakeholders; real estate and construction companies; agricultural entrepreneurs; authorities in charge for forests maintenance; national, regional and local authorities; social and environmental movements; associations; communities; and individuals. Other stakeholders support CMOs producers in the mitigation process. These include: (a) technology providers of any type; (b) private and public investment banks, as well as international and regional funds, in charge of providing

[58] See Sect. 6.4 of Chap. 6 of this book.

[59] See Sect. 6.5 of Chap. 6 of this book.

Table 8.3 Stakeholders involved in the supply of Certified Mitigation Outcomes

CMOs producers	Stakeholders supporting CMOs production	Certifiers, MRVers	Shareholders
Stakeholders reducing emissions	Technology providers	CMO Bank	Producers' clients, suppliers and competitors
Stakeholders avoiding emissions	Credit and financial support providers	Accredited auditing firms	Local communities
Stakeholders sequestrating emissions	Capacity builders	Non-governmental organisations	Individuals
	Consultants	Media	
		Consultants	

financial support to mitigation activities; (c) public and private stakeholders required to provide the necessary capacity for CMOs producers to act and to obtain corresponding certifications; (d) other consultants (i.e. legal and financial consultants) supporting CMOs producers.

Other stakeholders again are responsible for the MRV of the effectiveness of mitigation production. Some of these stakeholders shall operate under direct control of the CMO Bank, mandated to oversee the entire certification process within the MA. Others, such as non-governmental organisations, social and environmental movements, media, and consultants shall scrutinise the effectiveness and transparency of CMO, as well as the potentially adverse side effects resulting from mitigation activities, in principle in parallel to the CMO Bank MRV's initiatives. Local communities and individuals residing in areas affected by mitigation activities represent key stakeholders for MA mitigation supply's governance and, similarly to the CDM and the REDD+, these categories of stakeholders shall be involved in the process of preparing mitigation activities. Finally, changes in social and economic behaviours, such as those put in place by the CMOs producers through mitigating actions, may directly affect a range of traditional stakeholders, including producers' clients, suppliers and competitors.

8.11 Governing the Exchange of Mitigation Outcomes

Despite the CMO-EB being envisaged as the only authority in charge of the entire set of CMOs exchange options, the overall governance concerning these exchanges appears less obvious compared to the governance of demand and supply of mitigation outcomes. Due to a co-existence of approaches intended to facilitate the exchange of mitigation outcomes within the MA,[60] specifying a single governance system entailing the whole range of alternatives is virtually impossible. It is instead

[60] See Sect. 7.3 of Chap. 7 of this book.

necessary to take into account each of the approaches envisaged by the MA and to distinguish their individual governance characteristics. Section 7.3 of Chap. 7 of this book identified three main categories of approaches that are used here for the governance analysis: (a) market-oriented approaches associated with monetary exchanges; (b) hybrid approaches that may or may not imply monetary exchanges; (c) non-market based approaches excluding any form of monetary exchange. The first category of approaches can be further organised into three sub-categories, whose main stakeholders are summarised in Table 8.4.

Table 8.4 shows how the three sub-categories share at least seven involved stakeholders (the CMO-EB, the MA members, other stakeholders with quotas, CMO producers, law firms, accounting firms and MATF). While the CMO-EB and the accounting firms oversee and certify the correct implementation of CMOs exchanges envisaged by the sub-categories, some additional considerations concerning the remaining five stakeholders are required to fully understand the multi-layered system of market-oriented approaches associated with monetary exchanges.

MA members primarily act as final purchasers of CMOs in each of the three sub-categories, yet possibly may also act as CMOs producers when directly producing mitigation outcomes in the form of individual stakeholders and then selling CMOs through the listed sub-categories. Other stakeholders charged with quotas by MA members play a role very similar to the one played by the same MA members. They act primarily as final purchasers of CMOs[61] and possibly also as individual stakeholders producing and selling them. In principle, CMOs producers are present in all three sub-categories. An exception to the CMOs producers' constant presence may be envisaged in the case of markets that include only secondary exchanges. These types of exchanges may ensure that CMOs reach markets via intermediaries, thus excluding their producers from directly engaging in market' activities. According to this hypothesis the sub-category would not directly lead to a production of CMOs, with this being confined to a exchange occurring prior to the markets exchanges. By potentially providing support and assistance to any stakeholder involved in CMOs exchanges, law and consultancy firms are implicitly included in all sub-categories. So is the MATF, the final recipient of all CMOs.

Brokers, banks, investment firms, and other private and public investors represent the necessary financial intermediaries for the functioning of the first two sub-categories included in Table 8.4. For instance, if investments are provided through a repurchase agreement, they may directly support the production of CMOs, thus serving as 'loans' to be repaid in corresponding quantities of CMOs. If investments are provided once CMOs have been produced, intermediation by above-listed stakeholders may broker a deal between producers and investors. This second option would be beneficial especially for small-scale CMOs producers lacking the necessary capabilities and skills to engage in spot or derivate markets. Again, in this second case, brokers, banks, investment firms, and other private and public investors would *de facto* exclude CMOs producers from playing a direct role in the markets.

[61] In this process CMOs would be transferred one more time through a non-monetary exchange in favour of the MA member under whose domain the stakeholders operate.

While being present in virtually any option concerning the CMOs production, technology and service providers only exist as stakeholders in the exchange of CMOs if and when direct bilateral trade between suppliers and final users shifts from monetary to non-monetary exchange. In case of non-monetary exchanges,[62] involved stakeholders would exchange CMOs through services and/or products. Technology and capacity building for CMOs production may become privileged options for stakeholders operating in non-monetary exchanges. By envisaging in-kind payment for supplied services and/or products in the form of CMOs resulting from these services/products' application, this form of exchange appears to align with bilateral/multilateral agreements between MA members and/or other public stakeholders acting under the MA.

Table 8.5 summarises the hybrid approaches to CMOs exchanges under the MA, which may or may not imply monetary exchanges. These approaches are organised into four main sub-categories[63]: (a) monetary, mitigation-aimed taxation within the MA; (b) CMO-based taxation within the MA; (c) trade sanctions and border carbon adjustments; (d) share of proceeds.

The first sub-category (monetary, mitigation-aimed taxation within the MA) envisages MA members or other stakeholders entitled to apply taxation under their jurisdictions to raise funds for direct investment in mitigation activities by charging *ad hoc* fees in their domains. This kind of taxation does not provide an *a priori* quantification of CMOs to be produced. It is entirely focused on their production *per se*. Stakeholders involved in this sub-category include any form of stakeholder subject to taxation and CMOs producers benefitting from the collected funds. While correlated CMOs production dynamics may potentially imply the involvement of any stakeholder listed in Table 8.3, the authorities imposing taxation will be free to allocate the collected funds to mitigating sectors and projects of their choice. As a consequence, funds allocation can determine climate strategies and policies of these authorities and contribute to the implementation of their NDCs/MDCs.

Compared to the first sub-category, CMO-based taxation within the MA does not imply any direct support to *ad hoc* mitigating activities, instead requiring stakeholders subject to taxation to source required CMOs by using other approaches envisaged by the MA for CMOs exchanges. While MA members and other stakeholders entitled to apply taxation under their jurisdictions shall use this specific sub-category with the primary aim of collecting CMOs necessary to meet their quotas, they may nevertheless establish rules concerning the origins of the CMOs collected through taxation (i.e. limitations regarding specific sectors, exchange approaches, or designated geographic areas). This sub-category can therefore also be used as a tool for contributing to NDCs/MDCs.

The third sub-category mainly aims at hitting entities external to the MA system and imposing trading sanctions and/or Border Carbon Adjustments (BCA) on their products and services imported to the MA in case they do not abide by the MA principles and criteria. As discussed above in Sect. 8.3, their application may face

[62] See Sect. 7.3 of Chap. 7 of this book.

[63] See also Sect. 7.3 of Chap. 7 of this book.

Table 8.4 Stakeholders involved in market-oriented approaches associated with monetary exchanges

Sub-category	Involved stakeholders	Production of mitigation outcomes
Trading on a spot/derivate market for climate-related asset exchange	CMO-EB	Possibly
	MA members	
	Other stakeholders with quotas	
	CMOs producers	
	Banks	
	Investment firms	
	Brokerages	
	Other private and public investors	
	Law and consultancy firms	
	Accounting firms	
	MATF	
Transfer using a broker to privately and directly match suppliers and final users	CMO-EB	Yes
	MA members	
	Other stakeholders with quotas	
	CMOs producers	
	Banks	
	Investment firms	
	Brokerages	
	Other private and public investors	
	Law and consultancy firms	
	Accounting firms	
	MATF	
Direct bilateral trade between suppliers and final users	CMO-EB	Yes
	MA members	
	Other stakeholders with quotas	
	CMOs producers	
	Law and consultancy firms	
	Accounting firms	
	Technologies and services providers	
	MATF	

limits and constraints within the WTO and requires a careful analysis of the correlated legal risks. As a consequence, legal firms will play a significant role in supporting the definition of such trading sanctions. MA members will maintain full responsibility for the final destination of the funds collected through this sub-category, ideally allocating them to combat climate change (either in the form of further support for mitigating actions or as a tool for supporting adaptation and resilience). Thanks to this, the third sub-category may become another tool to contribute to NDCs/MDCs.

It is important to remember how calculations of the MA members' carbon emissions include the application of a non-monetary form of BCA, which directly affects MA members.[64] Absolute carbon emissions levels of MA members used to calculate their quotas will for instance include both carbon emissions within the members' borders and carbon emissions associated with the production of goods and services that, produced outside the MA domains, will be imported and used within the MA members' jurisdictions. By adopting this specific approach to BCA, which does not infringe any norm of the WTO, the MA virtually excludes any possible interest in outsourcing emissions beyond its domain, thereby avoiding carbon leakage.

Share of proceeds represents a somewhat different sub-category among hybrid approaches to CMOs exchange envisaged by the MA. When applied according to Paragraph 6 of A6PA, share of proceeds only covers MA administrative costs and supports adaptation policies aimed at least developed areas. Beneficiaries of this sub-category include all MA authorities and those stakeholders benefitting from the funded adaptation actions. In the absence of more specific criteria for allocating share of proceeds' funds, they may eventually be used to also support areas and stakeholders beyond the MA jurisdiction. Yet, MA members and other stakeholders entitled to apply taxation under their jurisdictions may decide to apply additional share of proceeds under their jurisdictions. In doing so they may eventually select other allocation criteria for collected funds, including mitigation actions support or adaptation and resilience interventions limited to their jurisdiction. As a consequence, such additional options for establishing share of proceeds may extend to a wide range of stakeholders.

In principle, revenues collected from tax-based approaches (sub-categories included in Table 8.5) may support institutional (public and private) climate funds, e.g., the Green Climate Fund (GCF). Tax-based approaches in any form can become significant triggers for initiatives encompassing large-scale programmes, focused on specific actions, sectors and/or geographic areas. If channelled towards such actions, any revenue derived from the sub-categories listed in Table 8.5 may contribute to significantly expanding the range of involved stakeholders. While it is not possible to suggest a complete list of potential beneficiaries, there is no doubt that discretion given to MA members and other stakeholders in allocating funds may provide them with additional support for their NDCs/MDCs.

To complete the review of stakeholders potentially involved in the process of CMOs exchange it is necessary to discuss two specific approaches already intro-

[64] See Sect. 5.3 of Chap. 5 of this book.

Table 8.5 Stakeholders involved in hybrid approaches

Sub-category	Involved Stakeholders	Production of mitigation outcomes
Monetary, mitigation-aimed taxation within the MA	CMO-EB	Yes
	MA members	
	Other stakeholders with quotas	
	Other stakeholders subject to taxation within the MA	
	Beneficiaries from collected funds (CMOs producers)	
	Accounting firms	
CMO-based taxation within the MA	CMO-EB	No
	MA members	
	Other stakeholders with quotas	
	CMOs producers subject to taxation	
	Accounting firms	
Trade sanctions and border carbon adjustments	CMO-EB	Possibly
	MA members	
	Non-MA members	
	Collected funds beneficiaries[a]	
	Law firms	
	Accounting firms	
Share of proceeds	CMO-EB and any other MA authorities	Possibly
	MA members	
	Other stakeholders with quotas	
	Beneficiaries from collected funds	
	Accounting firms	

[a]The existence of these stakeholders depends on the collected funds destination, which may or may not be of productive nature

duced in Sect. 7.3 of Chap. 7 of this book: (a) climate-related quantitative easing; (b) direct CMOs production. The term 'Green Quantitative Easing' was first adopted by Murphy and Hines (2010) and linked to climate change by a group of scholars (Hourcade et al. 2012; Rozenberg et al. 2013; Hourcade et al. 2014; Aglietta et al. 2015; Hourcade et al. 2015). Designed for application within the European Union (EU), climate-related quantitative easing suggests that central banks provide financial intermediaries with funds, granted by governments and addressed specifically to finance mitigation actions. This approach gained interest worldwide after the approval of the PA (Campiglio 2016; Dafermos et al. 2016; Global4C 2016). Stakeholders involved in MA quantitative easing would include: (a) governments of MA members acting as public authorities offering guarantees to the funds adopted for quantitative easing; (b) MA members' central banks acting as providers of funds; (c) investment banks and financial institutions acting as intermediaries between the central banks funds and CMOs producers; (d) CMOs producers as final beneficiaries of the funds.

Thanks to this approach, CMOs producers may obtain *ex ante* full support for the implementation of their mitigating actions, while MA members would radically cut costs in meeting their quotas. An interesting option may be its application in Sharia-based jurisdictions. For instance, by comparing any form of monetary interest to usury, Islamic law bans interest-based loans, or *riba*, within its jurisdiction. By exchanging a non-monetary form of compensation for loans, such as CMO, climate-related quantitative easing may be applied in Islamic jurisdictions without infringing fundamental religious principles.[65] However, climate-related quantitative easing may have significant inflationary impacts across economics adopting it.

As reported in Sect. 7.3 of Chap. 7 of this book, direct CMOs production represents the only wholly non-monetary approach to CMOs exchange currently envisaged within the MA. It is based on the simple logic that MA members with quotas can decide to meet (part of) these shares through direct mitigation activities. In this case, the number of involved stakeholders appears to be reduced in comparison to the other approaches. MA members act also as CMOs producers with no intermediary between them and the MATF regarding the CMOs' final banking. Still including the CMO Bank and accounting firms as stakeholders in charge of certifying MA members' mitigation outcomes, direct production may eventually include additional stakeholders whose role would be to support MA members in implementing their mitigating actions. Among these stakeholders it is possible to envisage technology providers, capacity builders, and various typologies of consultants.

Some final remarks to conclude this section concern the stakeholders involved in mitigation actions that do not obtain official certification through the MAM-based system envisaged by the MA. Analysed in Sect. 7.4 of Chap. 7 of this book, these types of mitigation still play a significant role within the overall MA architecture, although they do not count towards meeting the NZC target and its quotas distribution established by the MA. These activities modify absolute carbon emissions levels of MA members, hence impacting their positioning in the formula application. As a consequence, non-certified mitigation outcomes maintain a value within the MA. While MA structures have no authority on the management of non-certified mitigation, the variety of activities eventually belonging to this 'pseudo-approach'[66] may lead to an increase in the quantity and types of stakeholders linked to the MA far beyond the 'boundaries' established by the other approaches analysed here. While describing and analysing all the stakeholders eventually involved in the management of non-certified mitigation outcomes would be excessively long, if not

[65] Nevertheless Dafermos et al. (2016) emphasise how climate-related quantitative easing does not per-se turn out to be capable of granting quantified and substantial mitigation, hence requiring for other types of climate policies to be implemented in conjunction in order to meet the mitigation objectives posed by the PA. For this reason the MA includes climate-related quantitative easing as one of the complementary approaches to reach its ultimate goals (see also Sect. 7.3 of Chap. 7 of this book).

[66] See Sect. 7.4 of Chap. 7 of this book.

meaningless, for the purpose of this section it is nevertheless important to stress once more the importance of this 'pseudo-approach' in supporting MA members and other stakeholders in designing their climate policies and contributing to their NDCs/MDCs.

8.12 Conclusions: A New Climate Architecture

Despite their variety and differences, all elements and features analysed in this chapter eventually converge in defining the uniqueness and holistic harmony that characterise the MA governance in all its aspects. As such, this concludes the long description of the overall architecture composing the MA that started with Chap. 5 of this book, while at the same time encompassing all the key elements of the MA proposal' skeleton. In the description of stakeholders, as well as of networks that need establishing to meet the MA's goals and objectives, this Chapter summarises all the key elements characterising the MA. Demand, supply and exchange of mitigation outcomes are now fully explained along with their instruments, rules, scope, rationale, and embedded stakeholders.

The overarching, multi-level, hybrid model governing the MA has now been introduced. The size and shape of the MA described in the central four Chapters of this book is now concluded. The final part of this book, consisting of Chaps. 9, 10 and 11, outlines embedded potentials by providing an *a priori* analysis of the effects that the MA may have once fully implemented.

References

Aglietta, M., Hourcade, J.-C., Jaeger, C., & Peressin Fabert, B. (2015). Financing transition in an adverse context: Climate finance beyond carbon finance. *International Environmental Agreements, 15*, 403–420.

Andonova, L. B., Betsill, M. M., & Bulkeley, H. (2009). Transnational climate governance. *Global Environmental Politics, 9*(2), 52–73.

Andresen, S. (2015). International climate negotiations: Top-down, bottom-up or a combination of both? *The International Spectator, 50*(1), 15–30.

Bayer, P., Marcoux, C., & Urpelainen, J. (2013). Leveraging private capital for climate mitigation: Evidence from the clean development mechanism. *Ecological Economics, 96*, 14–24.

Bellassen, V., & Cochran, I. (2015). Introduction: Key notions and trade-offs involved in MRVing emissions. In V. Bellassen & N. Stephan (Eds.), *Accounting for carbon: Monitoring, reporting and verifying emissions in the climate economy* (pp. 1–17). Cambridge: Cambridge University Press.

Biermann, F., Pattberg, P., van Asselt, H., & Zelli, F. (2009). The fragmentation of global governance architectures: A framework for analysis. *Global Environmental Politics, 9*(1), 14–25.

Bodansky, D. (2016). The legal character of the Paris Agreement. *RECIEL, 25*(2), 142–150.

Bodansky, D. M., Hoedl, S. A., Metcalf, G. E., & Stavins, R. N. (2016). Facilitating linkage of climate policies through the Paris outcome. *Climate Policy, 16*(8), 956–972.

Böhm, S., Brei, V., & Dabhi, S. (2015). EDF Energy's green CSR claims examined: The follies of global carbon commodity chains. *Global Networks, 15*, 87–S107.

Böhmelt, T. (2013). Civil society lobbying and countries' climate change policies: A matching approach. *Climate Policy, 13*(6), 698–717.

Bradshaw, P. (2014). The dynamics of nested governance: A system perspective. In P. Bradshaw & M. Toubiana (Eds.), *Nonprofit governance: Innovative perspectives and approaches* (pp. 231–248). London: Routledge.

Buhr, K., Roth, S., & Stigson, P. (2014). Climate change politics through a global pledge-and-review regime: Positions among negotiators and stakeholders. *Sustainability, 6*, 794–811.

Campiglio, E. (2016). Beyond carbon pricing: The role of banking and monetary policy in financing the transition to a low-carbon economy. *Ecological Economics, 121*, 220–230.

Caparrós, A. (2016). The Paris Agreement as a step backward to gain momentum: Lessons from and for theory. *Revue d'Economie Politique, 126*(3), 347–356.

Chang, J.-P., & Bellassen, V. (2015). Trendsetter for territorial schemes: National GHG inventories under the UNFCCC. In V. Bellassen & N. Stephan (Eds.), *Accounting for carbon: Monitoring, reporting and verifying emissions in the climate economy* (pp. 21–71). Cambridge: Cambridge University Press.

Clémençon, R. (2016). The two sides of the Paris climate agreement: Dismal failure or historic breakthrough? *Journal of Environment and Development, 25*(1), 3–24.

Dafermos Y., Nikolaidi, M., & Galanis, G. (2016, November 14–15). Climate change, financial stability and monetary policy. Paper presented at the Bank of England workshop on central banking, climate change and environmental sustainability.

Das, K. (2015). Climate clubs: Carrots, sticks and more. *Economic & Political Weekly, 5*(34), 24–27.

De Coninck, H., Haake, F., & van der Linden, N. (2007). Technology transfer in the clean development mechanism. *Climate Policy, 7*, 444–456.

Deleuil, T. (2012). The common but differentiated responsibilities principle: Changes in continuity after the Durban conference of the parties. *RECIEL, 21*(3), 271–281.

Den Elzen, M. G. J., Schaeffer, M., & Lucas, P. L. (2005). Differentiating future commitments on the basis of countries' relative historical responsibility for climate change: Uncertainties in the "Brazilian proposal" in the context of a policy implementation. *Climatic Change, 71*(3), 277–301.

Doelle, M. (2004). Climate change and the WTO: Opportunities to motivate state action on climate change through the World Trade Organization. *RECIEL, 13*(1), 85–103.

Downie, C. (2014). Transnational actors in environmental politics: Strategies and influence in long negotiations. *Environmental Politics, 23*(3), 376–394.

Droege, S. (2011). Using border measures to address carbon flows. *Climate Policy, 11*, 1191–1201.

Du Pont, Y. R., Jeffery, M. L., Gütschow, J., Christoff, P., & Meinshausen, M. (2016). National contributions for decarbonizing the world economy in line with the G7 agreement. *Environmental Research Letters, 11*, 054005.

Falkner, R. (2016). A minilateral solution for global climate change? On bargaining efficiency, club benefits and international legitimacy. *Perspectives on Politics, 14*(1), 87–101.

Gallaraga, I., Gonzalez-Eguino, M., & Markandya, A. (2011). The role of regional governments in climate change policy. *Environmental Policy and Governance, 21*, 164–182.

Gandenberger, C., Bodenheimer, M., Schleich, J., Orzanna, R., & Macht, L. (2015). Factors driving international technology transfer: Empirical insights from a CDM project survey. *Climate Policy, 16*(8), 1065–1084.

Garibaldi, J. A. (2014). The economics of boldness: Equity, action, and hope. *Climate Policy, 14*(1), 82–101.

Geden, O. (2015, December 14). Paris climate deal: The trouble with targetism. *The Guardian*.

Global4C. (2016, May 2). Michael Metcalfe advocates Quantitative Easing (QE) for climate change mitigations. *Complementary Currencies for Climate Change Article*.

Goldemberg, J., & Guardabassi, P. (2015). Burden sharing in the implementation of the climate convention. *Energy Policy, 81*, 56–60.

Green, J. F., Sterner, T., & Wagner, G. (2014). A balance of bottom-up and top-down in linking climate policies. *Nature Climate Change, 4*, 1064–1067.

Haites, E., Duan, M., & Seres, S. (2006). Technology transfer by CDM projects. *Climate Policy, 6*, 327–344.

Hale, T. (2016). "All hands on deck": The Paris agreement and nonstate climate action. *Global Environmental Politics, 16*(3), 12–22.

Hare, W., Stockwell, C., Flachsland, C., & Oberthur, S. (2010). The architecture of the global climate regime: A top-down perspective. *Climate Policy, 10*, 600–614.

Helm, D., Hepburn, C., & Ruta, G. (2012). Trade climate change and the political game theory of border adjustments. *Oxford Review of Economic Policy, 28*, 368–394.

Hermwille, L. (2016). Climate change as a transformation challenge. A new climate policy paradigm? *GAIA, 25*(1), 19–22.

Hourcade, J. C., Perrissin Fabert, B., & Rozenberg, J. (2012). Venturing into uncharted financial waters: An essay on climate-friendly finance. *International Environment Agreements Politics, Law and Economics, 12*(2), 165–186.

Hourcade, J-C., Aglietta, M., & Perrissin Fabert, B. (2014). *Transition to a low-carbon society and sustainable economic recovery, A monetary-based financial device*. CIRED concept note.

Hourcade, J-C., Aglietta, M., & Perrissin Fabert, B. (2015). The low carbon transition between the animal spirits of finance and the fault lines of the world economy. In A. Sirkis (Ed.), *Moving the trillions – A debate on positive pricing of mitigation actions* (pp. 22–36). Rio de Janeiro: Brasil No Clima.

ICAO. (2016). *On Board a Sustainable Future*. 2016 Environmental Report, ICAO.

Jacobs, M. (2016). High pressure for low emissions. How civil society created the Paris climate agreement. *Juncture, 22*(4), 314–323.

Johnston, C., Milman, O., & Vidal, J. (2016, October 2016). Climate change: Global deal reached to limit use of hydrofluorocarbons. *The Guardian*.

Kashwan, P., & Holahan, R. (2014). Nested governance for effective REDD+: Institutional and political arguments. *International Journal of the Commons, 8*(2), 554–575.

Knox, J. (2004). The judicial resolution of conflicts between trade and the environment. *Harvard Environmental Law Review, 28*(1), 1–78.

Kuchler, M. (2015). Stakeholding as sorting of actors into categories: Implications for civil society participation in the CDM. *International Environmental Agreements*. doi:10.1007/s10784-015-9314-5.

Levi, M. D. (1991). Bretton woods: Blueprint for a greenhouse gas agreement. *Ecological Economics, 4*, 253–267.

Lund, E. (2013). Environmental diplomacy: Comparing the influence of business and environmental NGOs in negotiations on reform of the clean development mechanism. *Environmental Politics, 22*(5), 739–759.

Madden, N. J. (2014). Greenmeans stop: Veto players and their impact on climate-change policy outputs. *Environmental Politics, 23*(4), 570–589.

Mannis, A. (2001). *Indicators of sustainable development*. GAIA Report.

Marcu, A. (2016, January). *Carbon market provisions in the Paris agreement (article 6)* (CEPS Special Report, 128).

Maruyama, W. H. (2011). Climate change and the WTO: Cap and trade versus carbon tax? *Journal of World Trade, 45*(4), 679–726.

Mattoo, A., & Subramanian, A. (2012). Equity in climate change: An analytical review. *World Development, 40*(6), 1083–1097.

Mavroidis, P. C., & de Melo, J. (2015). Climate change policies and the WTO: Greening the GATT, revisited. In S. Barret, C. Carracro, & J. de Melo (Eds.), *Towards a workable and effective climate regime* (pp. 225–236). London: CEPR Press.

Methmann, C. (2013). The sky is the limit: Global warming as global governmentality. *European Journal of International Relations, 19*(1), 69–91.

Michaelowa, A. (2015). Opportunities for and alternatives to global climate regimes post-Kyoto. *Annual Review of Environment and Resources, 40*, 395–417.
Michaelowa, A., & Michaelowa, K. (2015). Do rapidly developing countries take up new responsibilities for climate change mitigation? *Climatic Change, 133*, 499–510.
Morgan, J. (2016). Paris COP 21: Power that speaks the truth? *Globalizations*. doi:10.1080/14747731.2016.1163863.
Müller, B., Höhne, N., & Ellermann, C. (2010). Differentiating (historic) responsibilities for climate change. *Climate Policy, 9*, 593–611.
Murphy, R., & Hines, C. (2010). Green quantitative easing: Paying for the economy we need. *Finance For the Future Report*.
Nordhaus, W. (2015). Climate clubs: Overcoming free-riding in international climate policy. *American Economic Review, 105*(4), 1339–1370.
Peters, G.P., Andrew, R.M., Solomon, S., & Friedlingstein, P. (2015). Measuring a fair and ambitious climate agreement using cumulative emissions. *Environmental Research Letters, 10*, 105004.
Pettenella, D., & Brotto, L. (2012). Governance features for successful REDD+ projects organization. *Forest Policy and Economics, 18*, 46–52.
Potoski, M. (2015). Green clubs in building block climate change regimes. *Climatic Change*. doi:10.1007/s10584-015-1517-9.
Rajamani, L. (2016). Ambition and differentiation in the 2015 Paris Agreement: Interpretative possibilities and underlying politics. *International and Comparative Law Quarterly, 65*, 493–514.
Raman, M. (2016, January 9). Climate change battle in Paris. *Economic and Political Weekly, 51*(2), 35–40.
Rogelj, J., den Elzen, M., Höhne, N., Fransen, T., Fekete, H., Winkler, H., Scheffer, R., Sha, F., Riahi, K., & Meinshausen, M. (2016). Paris Agreement climate proposals need a boost to keep warming well below 2 °C. *Nature, 354*, 631–639.
Rozenberg, J., Hallegatte, S., Perrissin-Fabert, B., & Hourcade, J.-C. (2013). Funding low-carbon investments in the absence of a carbon tax. *Climate Policy, 13*(1), 134–141.
Sabel, C. F., & Victor, D. G. (2015). Governing global problems under uncertainty: Making bottom-up climate policy work. *Climatic Change*. doi:10.1007/s10584-015-1507-y.
Shiroyama, H., Yarime, M., Matsuo, M., Schoreder, H., Scholz, R., & Ulrich, A. E. (2012). Governance for sustainability: Knowledge integration and multi-actor dimensions in risk management. *Sustainability Science, 7*(1), 45–55.
Stewart, R., Oppenheimer, M., & Rudyk, B. (2013). A new strategy for global climate protection. *Climatic Change, 120*, 1–12.
Streck, C. (2004). New partnerships in global environmental policy: The clean development mechanism. *Journal of Environment and Development, 13*(3), 295–322.
Streck, C. (2012). Innovativeness and paralysis in international climate policy. *Transnational Environmental Law, 1*(1), 137–152.
Thomson, M. C., Baruah, M., & Carr, E. R. (2011). Seeing REDD+ as a project of environmental governance. *Environmental Science and Policy, 14*, 100–110.
Van der Gaast, W., Begg, K., & Flamos, A. (2009). Promoting sustainable energy technology transfers to developing countries through the CDM. *Applied Energy, 86*, 230–236.
Vant, A., & Vedeld, P. O. (2013). National governance structures for REDD+. *Global Environmental Change, 23*, 422–432.
Victor, D. G. (2015, January). *The case for climate clubs*. E15 Initiative Thinking Piece.
Wara, M. (2008). Measuring the clean development Mechanism's performance and potential. *UCLA Law Review, 55*, 1759–1803.
Watson, J., Ockwell, D., Byrne, R., & Stua, M. (2015). Lessons from China: Building technological capabilities for low carbon technology transfer and development. *Climatic Change, 131*(3), 387–399.

Wei, Y.-M., Wang, L., Liao, H., Wang, K., Murty, T., & Yan, J. (2014). Responsibility accounting in carbon allocation: A global perspective. *Applied Energy, 130*, 122–133.

Weischer, L., Morgan, J., & Patel, M. (2012). Climate clubs: Can small groups of countries make a big difference in addressing climate change? *RECIEL, 21*(3), 177–192.

Well, M., & Carrapatoso, A. (2016). REDD+ finance: Policy making in the context of fragmented institutions. *Climate Policy*. doi:10.1080/14693062.2016.1202096.

Winkler, H., & Rajamani, L. (2014). CBDR&RC in a regime applicable to all. *Climate Policy, 14*(1), 102–121.

Wyborn, C., & Bixler, R. P. (2013). Collaboration and nested environmental governance: Scale dependency, scale framing, and cross-scale interactions in collaborative conservation. *Journal of Environmental Management, 123*, 58–67.

Zavodov, K. (2012). Renewable energy investment and the clean development mechanism. *Energy Policy, 40*, 81–89.

Zelli, F., & van Asselt, H. (2010). The overlap between the UN climate regime and the World Trade Organization: Lessons for climate governance beyond 2012. In F. Biermann, P. Pattberg, & F. Zelli (Eds.), *Global climate governance beyond 2012: Architecture, agency and adaptation* (pp. 79–96). Cambridge: Cambridge University Press.

Zhao, J. (2013). Developed countries' cap-and-trade border measures: China's possible reactions. *Chinese Journal of International Law, 12*(4), 809–842.

Part III
Effects of the Mitigation Alliance

Chapter 9
The Carbon Pricing and the Establishment of a Low Carbon Bretton Woods

Abstract Financing carbon emission mitigation in line with the general objectives of the Paris Agreement probably represents the largest challenge to its implementation. Scholars and policy makers have identified *carbon pricing* as a possible solution to this challenge. However, several issues, such as carbon price instability and its excessively low prices, still need to be solved. When discussing the Mitigation Alliance (MA)'s approach to *carbon pricing*, in this chapter emphasises how it differs from the traditional approaches by shifting *carbon pricing* from a cost perspective based upon carbon emissions to a value perspective based upon carbon mitigation. Thanks to this approach, termed Positive Carbon Pricing (PCP), in this chapter demonstrates how the MA can effectively trigger a redirection of funding towards efforts necessary to meet the MA Net Zero Carbon (NZC) target. This process also represents the first benefit provided by the MA exclusively to its members. Specifically, the Chapter emphasises the role of Certified Mitigation Outcomes (CMO) in conferring effective value on documented mitigation. It is this value base which establishes CMO as the *reference currency* for carbon. Following on from the links between carbon as a *commodity* and CMO it is possible to theorise a MA monetary system that, due to its features, may be defined as Low Carbon Bretton Woods.

Keywords Carbon • Commodity • Carbon pricing • Positive Carbon Pricing (PCP) • Representative currency • Low Carbon Bretton Woods

9.1 Introduction

There is no doubt that, if implemented, the primary impact of the MA proposal will be aggregate, net mitigation of carbon[1] emissions (or simply mitigation). The whole structure and functioning of the MA as described in the previous Chapters of this book converge towards this mitigation effect. If efficiently implemented, the MA will lead to real, quantifiable and consistent mitigation in line with the primary objectives defined by the Paris Agreement (PA). Nevertheless, these outcomes

[1] Carbon Dioxide (CO_2) is the most common Greenhouse Gas (GHG) and is often used as a reference unit to express levels of GHG emissions (expressed in CO_2 equivalent, or CO_2eq). In the context of this Chapter and, more in general, of this book the term carbon, expressing CO_2, and the term GHG are synonyms, hence carbon is used as single term to express GHG.

M. Stua, *From the Paris Agreement to a Low-Carbon Bretton Woods*,
DOI 10.1007/978-3-319-54699-5_9

remain theoretical in the absence of the funds necessary to pay for targeted mitigation.

As described in Chap. 3 above, aggregate carbon qualifies as a non-excludable, public good with benefits not limited to actors involved in mitigation. The MA requires both commitment of the largest possible number of stakeholders to meet its mitigation target and the corresponding financial resources to fund these efforts (McCollum et al. 2013). At the same time, it is obvious that a MA limited to the production of non-excludable benefits would neither be able to commit its members to the common mitigation effort nor would it attract the necessary financial flows, resulting in a systemic failure. Like any form of transformational club,[2] the MA therefore requires private, excludable benefits for its members which reward and justify commitment and involvement while simultaneously stimulating the flow of the necessary private and public finance.

As described in Sect. 3.4 of Chap. 3 of this book, there are three systems to developing excludable benefits to promote mitigation: (a) an indirect system, where benefits for club members are not climate-related and mitigation happens as a side-effect; (b) a direct system, where mitigation is itself transformed into a private good; and (c) a hybrid system, which combines elements from the other two options. The MA represents such a hybrid system.

This Chapter and Chap. 10 demonstrate the ability of the MA's hybrid system to attribute excludability to benefits linked to mitigation efforts. While Chap. 10 focuses on indirect benefits, this chapter focuses on direct benefits by explaining how the MA can transform a public, non-excludable good, such as carbon, into a private, excludable good in order to raise the funds required to finance MA members' and stakeholders' mitigation efforts. The ability to attract private and public investment (Campiglio 2016), amounting to trillions of American dollars (Sirkis 2016), is essential for effective climate governance.

In order to clarify the direct benefits of the MA, this Chapter refers back to some of the most significant passages of previous Chapters. On the basis of a defined Net Zero Carbon (NZC) target that translates carbon into an exhaustible and quantified *commodity* for MA members, Sect. 9.2 describes how Certified Mitigation Outcomes (CMOs) acquire a value, which correlates to the obligations of MA members to meet their assigned shares of the mitigation burden or quota.[3] MA members would thus jointly meet the aggregate target by banking a corresponding quantity of CMOs. This intrinsic link between the MA mitigation target and CMO is the essential prerequisite for converting carbon into a private, excludable good within the MA.

With carbon becoming a private excludable good, their conservation expressed in CMOs implicitly carries a value. In describing the value base of CMOs within the MA, this Chapter taps into the wider debate on the rationale and use of *carbon pricing*. Section 9.3 focuses on *carbon pricing* in the context of the MA and defines

[2] See Chaps. 3 and 4 of this book.

[3] Quota is used in this Chapter and more in general in this book as synonym of 'assigned share of mitigation burden'.

Positive Carbon Pricing (PCP) as an alternative to conventional *carbon pricing* models.

Reinterpreting carbon as exhaustible and quantifiable and establishing PCP on the basis of the value of CMO leads to another significant advance in defining the direct, excludable benefits for MA members. This interlinked process transforms quantified and exhaustible carbon into a *commodity* for MA members and CMO into its sole *representative currency*. As money used within the MA needs to be converted into CMOs to allow members to meet their quotas, CMOs themselves hold a fixed exchange rate towards their *commodity* (carbon). This concept, with special reference to the potential effects on monetary economics, is discussed in Sect. 9.4. Drawing on the 1991 paper by Maurice David Levi titled "Bretton Woods: Blueprint for a greenhouse gas agreement" (Levi 1991) as its founding pillar, Sect. 9.4 proposes the establishment of a Low Carbon Bretton Woods with CMO as the *representative currency* of the MA.

The concluding Sect. 9.5 summarises the findings of this chapter and how they support the concept of benefits within the MA. Once more, it is pointed out how elements comprising the MA hybrid governance model cannot be understood as separate entities, but that instead they represent interlinked aspects of a single, multi-level, harmonious framework.

9.2 Privatising Carbon as a Consequence of the Founding of the Mitigation Alliance

Possible interactions between the MA and the World Trade Organization (WTO) jurisdiction[4] are the starting point for understanding how, within the MA, carbon represents a 'privatised resource' as opposed to a non-excludable, public good. Hence, Sect. 9.2 starts with a brief review of the findings of Chap. 8 in order to cast light on this privatisation process.

To date, international law has not yet fully addressed potential conflicts between the international legal regime on climate change, represented by the United Nations Framework Convention on Climate Change (UNFCCC), and the international legal regime on trade (the WTO). Strategies and policies required to deal with climate change may infringe existing international trade regulations established by the WTO's General Agreement on Tariffs and Trade (GATT). According to current international legislation, only exceptions allowed for by the GATT offer opportunities for implementing climate-related actions. GATT Article XX determines that two types of environment-oriented measures that, if applied in conjunction with restrictions on domestic production or consumption, in principle do not infringe the WTO principles. The first exception (GATT Article XX Paragraph (b)) refers to measures necessary to protect human, animal or plant life, or health. The second

[4] See Sect. 8.2 of Chap. 8 of this book.

(GATT Article XX Paragraph (g)) concerns measures relating to the conservation of exhaustible natural resources.

At first glance, the exception described in GATT Article XX Paragraph (b), which explicitly addresses adverse effects of anthropogenic actions, may more easily apply to the climate change context. Nevertheless, as no clear, specific and quantifiable definition of the ecological damage of climate change on humans, animals and plants has been agreed on at global level, measures seemingly covered by this Paragraph may be contested within the WTO regime. As a consequence, Paragraph (g) of GATT Article XX appears to be the only viable tool for establishing a set of norms to safeguard the MA's climate mitigation strategies and, with them, MA members' exclusive access to associated excludable benefits. Weischer et al. (2012, p. 188) stress that "if the environmental exception contained in GATT Article XX(g) were to be used to justify differences in treatment, it would have to be shown that the measure closely relates to the environmental objective". This specification serves as the rationale for applying the exception in question to the MA.

The rationale for the application of the GATT "environmental objective" exception is founded upon a new mind-set towards carbon. Commonly perceived as a non-exhaustible natural resource whose consumption equates with its emissions, carbon becomes an exhaustible resource when its permissible emission is capped under a maximum peak or ceiling. By defining an aggregate peak corresponding to a specific volume of carbon emissions at the beginning of the first timetable,[5] the MA establishes carbon as an exhaustible resource within its domain.[6] This step follows the pathway suggested by McKinnon (2015).

If carbon emissions correspond to the consumption of an exhaustible natural resource (carbon), mitigation equates to the conservation of the resource, consistent with the environmental exception contained in GATT Article XX(g). In designing a NZC mitigation pathway, the MA defines its aggregate mitigation target in line with the general objectives of the Paris Agreement (PA). The concrete measures ensuring the MA meets its aggregate mitigation target thus represent the "environmental objective" required to be acknowledged as an exception under the GATT (WTO) rules.

As described in Chaps. 5, 6, 7, 8 of this book, measures associated with the structure and functioning of the MA are components of a single governance model, and aim to determine and quantify the MA's environmental objective, which is to meet its NZC aggregate emission target by the end of this century. As all these structural and functional measures are interlinked because of the multi-level, hybrid nature of

[5] See Chap. 5 of this book.

[6] Interpreting carbon as an exhaustible resource (according to GATT/WTO) requires the establishment of a carbon budget. The carbon budget refers to the maximum amount of carbon that may be released to ensure that, with some probability, global average temperature will not exceed a specified limit. For example, if want to maintain a > 66% probability of limiting global average temperature rise to 2 °C above pre-industrial levels, the planet can emit an additional 1000Gt of carbon until 2100 (IPCC 2014; Zaman et al. 2016). This 1000Gt of expendable carbon is thus an exhaustible natural resource, i.e. a private good.

the MA governance structure,[7] they all have to be potentially adoptable in order to guarantee that the MA's overall environmental objective is met. As a consequence, any attempt to block the adoption of one or more of such measures by referring to international trade law would be illegal and thus must fail due to the exceptions granted by GATT Article XX(g).

The Mitigation Alliance Mechanism (MAM), by means of which MA members meet their quotas during the timetables, is the tool to be used by the MA to reach its environmental objectives. As mitigation equates to carbon conservation, CMOs issued through the MAM represent the only applicable unit of measurement for quantifying the MA's carbon conservation objectives, as expressed by its mitigation target.

The quantifiable, exhaustible nature of carbon, coupled with CMO exchangeability through the multiple approaches described in Chap. 7 of this book, transform CMO into a *commodity* within the domain of the MA. While the MA mitigation target distributed over time represents the floor quantity of the *commodity* demand, carbon emission represents the consumption of the *commodity* and mitigation the conservation of the *commodity*. The interpretation of carbon as a *commodity* is not an innovation of this book; rather, the concept is implicit in *carbon pricing* proposals generally

9.3 Carbon Pricing in Light of the Paris Agreement and of the Mitigation Alliance

The exclusion of ecological goods and services from the market pricing system is the most significant cause of the existing gap between the financial cost of a transition to an economic system in line with the PA's objectives and actual fund-raising provisions (Campiglio 2016). Scholars and policy-makers have identified an opportunity for bridging this gap in *carbon pricing*. In a broad sense, *carbon pricing* traditionally refers to any economic instrument capable of capturing what are known as external costs (or externalities) of carbon emissions[8] and tying them to their sources through a price. As *carbon pricing* refers to the external cost of carbon emissions (or carbon consumption), mitigation outcomes, which represent carbon conservation, refer to the carbon value. As already discussed in Sect. 9.2, recognising a value in carbon equates to recognising carbon as a *commodity*.

By internalising costs of external carbon emissions, *carbon pricing* simplifies investments aimed at addressing these costs, hence favouring the flow of investments towards mitigation actions. Ultimately, *carbon pricing* acts as a unit of measurement for investments made possible by the mitigation outcomes *per se*, hence becoming a tool to determine the net value of mitigation-related investments. As a

[7] See Sect. 8.2 of Chap. 8 of this book.

[8] Costs not directly paid by the carbon emitters.

consequence, the MA requires its own *carbon pricing* in order to assign a value to its *commodity* supply unit (the CMO). To understand this process, Sect. 9.3 first summarises state-of-the-art theoretical and empirical *carbon pricing* and describes its functioning, current achievements, and main shortcomings, before focusing on MA *carbon pricing*. This section demonstrates how most of the current *carbon pricing* shortfalls can be overcome through the MA.

Emissions Trading Systems (ETS) and carbon taxation represent the two commonly recognised instruments to establish *carbon pricing* (Tietenberg 2013; Jenkins 2014).[9] The ETS establishes a market price for carbon emissions by creating supply and demand for emission allowances (flexible *carbon pricing*)[10] and eventually by fixing the carbon price, defined as floor price, through auction sales of allowances. As well as determining a carbon price, the ETS ensures that envisaged mitigation outcomes actually occur by keeping aggregate emissions within their pre-allocated carbon budget,[11] which corresponds to the aggregate allowances.

Carbon taxation is a form of *carbon pricing* because it taxes emissions (or the carbon content of fossil fuels). Compared to the ETS, carbon taxation pre-defines *carbon pricing* but does not predetermine mitigation outcomes.[12] Despite the two approaches being generally considered as alternatives, Gollier and Tirole (2015) envisage their co-existence, as this would help to balance the inevitable trade-offs of the two forms of *carbon pricing* in operation side-by-side.

A recent World Bank report on the subject (World Bank 2016) provides contrasting data on the current status of *carbon pricing* worldwide. Encouraging information includes:

- (a) about 100 parties, accounting for almost 60% of global carbon emissions, stated interest in adopting *carbon pricing* in their (Intended) Nationally Determined Contributions ((I)NDCs) submitted to the UNFCCC[13];
- (b) initiatives for common *carbon pricing* standards are now involving coalitions of parties and other stakeholders, such as the Partnership for Market Readiness, Networked Carbon Markets and the Ministerial Declaration on Carbon Pricing;
- (c) on a regional, national and subnational level, about 40 national jurisdictions and over 20 cities, states, and regions placed a price on carbon in 2016:
- (d) over 1200 companies were using an internal carbon price in 2016; and
- (e) the level of carbon emissions covered by *carbon pricing* has increased threefold over the past decade.

---

[9] Yet Aldy and Stavins (2012) include emissions reduction credits in the list of the possible pricing paths.

[10] For an overview of the rationale, structure, and functioning of ETS see Sect. 7.2 of Chap. 7 of this book.

[11] See note 6.

[12] For an overview of the rationale, structure, and functioning of the various forms of carbon taxation see Sect. 7.3 of Chap. 7 of this book.

[13] Further details on (I)NDCs can be found at: http://unfccc.int/focus/indc_portal/items/8766.php.

Discouraging information suggests that during 2015 carbon prices ranged from less than US\$1/tCO_2eq to US\$131/tCO_2eq and that about three quarters of emissions covered were priced at less than US\$10/tCO_2eq.[14] The World Bank report stresses how the two major problems affecting *carbon pricing*, namely excessively high variability and low value, remain unsolved. As for the first problem, ETS fragmentation may cause significant variability in *carbon pricing* across regions, and this in turn is associated with limitations to competitiveness, welfare losses and carbon leakage (Lanzi et al. 2012). Similar issues may also afflict a fragmented regime of carbon taxation, with firms operating under jurisdictions that impose carbon taxes being heavily affected by associated costs.

While attempts to make exceptions for emission-intensive industries may lead to a race to the bottom in respect of both the allowances auctions and the levels of carbon-related taxation, a consistent global carbon tax has been identified as the only vehicle capable of ensuring that *carbon pricing* becomes an effective tool to mitigate carbon emissions (Stiglitz 2015) and to develop a global carbon market (Matthews 2007). Although it is the most logical solution to guarantee *carbon pricing* consistent with the PA objectives, the adoption of a global carbon tax of US\$100/tCO_2eq as proposed by Stiglitz (2015) appears highly unfeasible due to the multitude of political, economic and social agents who are unwilling to pay directly for mitigation (Jenkins 2014).

Limits posed by the current interpretation of *carbon pricing* led to a call for new, broader approaches towards recasting *carbon pricing* as an effective tool to promote carbon mitigation (Aglietta et al. 2015; Campiglio 2016). Bearing in mind that the MA proposes a variety of approaches for CMO exchange, including ETS and carbon taxation,[15] it is now possible to identify what may be the ideal MA *carbon pricing* system and to assess whether a *carbon pricing* reform may be consistent with the MA proposal.

In order to fully understand the MA *carbon pricing* process, it is first necessary to identify features that set it apart from the traditional ETS and carbon taxation. Compared to carbon taxation, MA *carbon pricing* does not determine a fixed monetary price for carbon. Instead, MA *carbon pricing* determines a non-monetary, fixed ceiling of carbon resource (the mitigation outcomes) which has to be safeguarded within a time limit (the MA target) and which is organised into similarly fixed, floor quantities of mitigation sub-targets to be met within specific timetables.[16] Compared to carbon taxation,[17] the MA has the significant advantage of guaranteeing a predetermined amount of mitigation, therefore leading to a defined mitigation pathway in line with the PA objectives. Yet, while carbon taxation offers

[14] Far below the price level required to effectively pursue at least the 2°C objective of the PA, estimated to be between US\$ 80/tCO_2eq and US\$120/tCO_2eq by 2030 (IPCC 2014; IEA 2015).

[15] See Sect. 7.3 of Chap. 7 of this book.

[16] The burden sharing defined by the application of the formula introduced in Sect. 5.3 of Chap. 5 of this book.

[17] While establishing a fixed price to be paid for each predetermined unit of carbon emission, carbon taxation cannot guarantee a link between the tax revenue and mitigation outcomes.

certainty regarding the *carbon pricing*, the MA *carbon pricing* appears to lack this certainty. To understand how the MA overcomes this uncertainty, its *carbon pricing* must be compared with the ETS *carbon pricing* system.

Although the MA *carbon pricing* may appear similar to the ETS *carbon pricing* system, where quantities of allowances to emit are predetermined and distributed among the relevant actors, there are important differences. The ETS is based on fixed quantities of emission liabilities (*carbon* emissions that in principle represent externalities or external costs, as discussed above) and the determination of an associated carbon price. MA *carbon pricing* is based on demand for specific assets (mitigation outcomes in the form of CMOs). Following the logic of the *commodity* interpretation of carbon described in Sect. 9.2, ETS *carbon pricing* depends on the cost of *commodity* consumption (carbon emission), whereas MA *carbon pricing* depends on the value of *commodity* conservation (carbon mitigation). In other words, while the ETS system prejudices *carbon pricing* with the negative connotation of a liability, the MA proposes an asset-based, Positive Carbon Pricing (PCP) founded on the intrinsic value of mitigation outcomes.

Understanding the rationale for PCP is a precondition for grasping its potential as an instrument for quantifying investments for mitigation in line with the PA objectives. The PCP concept was introduced during the Rio Clima - Rio Climate Challenge Conference promoted by Alfredo Sirkis[18] in 2012 (Pagina22 2013), initially as an alternative to ETS and carbon taxation for the purpose of boosting finance for mitigation actions and effective decarbonisation worldwide (Sirkis 2016).

With this aim in mind, PCP promoters attempted to shift from the traditional, cost-based view of carbon emissions, perceived as economic liabilities since the publication of the Stern Review (Stern 2006), to a new, positive view of the mitigation of carbon emissions, now cast as assets which counterbalance the negative externalities (Dasgupta 2016; Hourcade et al. 2016). PCP was therefore designed as the instrument to measure the intrinsic values of mitigation outcomes as assets to offset the external cost of carbon emissions (Stua et al. 2016).

The November 2015 Brazilian Submission to UNFCCC[19] proposed that PCP be recognised formally. It was eventually included in the PA, which states that the Conference of Parties serving as the meeting of the Parties to the Paris Agreement (CMA) "recognizes the social, economic and environmental value of voluntary mitigation actions and their co-benefits for adaptation, health and sustainable development".[20]

With the official recognition of the Social, Economic and Environmental Value of Mitigation Actions (SEVMA), PCP can now be used to assess the value of mitigation, hence becoming a unit of measurement for mitigation-related investments.

[18] At that time, Alfredo Sirkis was a Brazilian congressman who became President of the Brazilian Parliamentary Commission on Climate Change (CMMC) two years later.

[19] The same submission played a significant role during the whole process of developing PA Article 6 (Marcu 2016).

[20] PA Section IV, Paragraph 108 (UNFCCC 2015).

More specifically, adopting PCP implies that mitigation outcomes are assigned a SEVMA as a result of carbon conservation in its *commodity* form. By coupling the value of mitigation outcomes expressed in PCP with the system governing mitigation outcome demand, supply and exchange as described in Chaps. 5, 6, 7, 8 of this book, the MA transforms mitigation outcomes into valuable financial assets. As a consequence of this positive valuation of *carbon* as a resource, associated investments will surely stimulate mitigation-related reinvestments.

Furthermore, by establishing CMO as the only recognised mitigation-related asset within its regime and by establishing a floor quantity of CMOs demand (represented by the MA NZC target and its spread over the different timetables), the MA transforms CMO from an asset into a currency, which is used by its members to acquire specified quantities of mitigations. As the only instrument available for MA members to demonstrate their achievement of their quotas CMO functions as the only unit of measurement for mitigation outcomes in operation within the MA. As mitigation outcomes represent the conservation of carbon in its *commodity* form, CMO thus becomes the carbon *representative currency* within the MA.

Using CMO as the *representative currency* for carbon automatically implies the need to establish a fixed exchange rate between the *representative currency* and its reference *commodity*. Once established, the MA shall fix the exchange rate between CMO and its reference *commodity* (carbon) in a way similar to its main predecessor, the Clean Development Mechanism (CDM),[21] with its Certified Emissions Reductions (CERs), each unit of which equalled 1 metric tonne of CO_2 equivalent ($MtCO_2eq$). When applied to CMOs, the respective equivalence is the fixed exchange rate between CMO and carbon, its corresponding *commodity*. With the MA determining CMO as sole tool for converting actual mitigation outcomes into assets exchangeable among the MA members, CMO becomes the *representative currency* for carbon, hence expressing its effective value. This means that the CMO value represents the fixed price of carbon as a *commodity*. The price paid by MA members and stakeholders for purchasing a single unit of CMO through any exchange approach described in Sect. 7.3 of Chap. 7 of this book thus represents the flexible exchange rate between CMO as the MA's *commodity* currency and any MA member's national currency, and ultimately functions as the actual *carbon pricing* inside the MA.

Playing a role similar to the one played by the US$ in the Bretton Woods system,[22] CMO would then become a *representative currency* for all MA members. This fundamental innovation introduced by the MA could lead to effects that go far beyond the climate change domain, possibly even helping to overcome systemic failures that have afflicted the international monetary system since the advent of fiat

[21] See also Chap. 6 of this book.

[22] Bretton Woods established a fixed exchange rate between U$D and gold (U$D35 per ounce of gold). The flexible exchange rate between the US dollar and any other currency linked to the system allowed for the linked jurisdictions to constantly check the status of their economic wealth by indirectly comparing it to gold.

currencies. Due to its significance and potential, the CMO system initiated by the MA is hereby defined as the Low Carbon Bretton Woods.

9.4 The Low Carbon Bretton Woods

More than 25 years ago, 1 year before the 1992 Rio Earth Summit that established the UNFCCC, Maurice David Levi published a paper titled: "Bretton Woods: Blueprint for a greenhouse gas agreement" (Levi 1991). In that paper, Levi identified several similarities between the international regime embodied by Bretton Woods and the regime emerging around climate change challenge and formulated hypothetical framework for a Bretton-Woods-style carbon governance regime.

To Levi, aiming at a public non-excludable good represented the key analogy between Bretton Woods (designed for the post-war reconstruction) and a necessary climate regime (where carbon represents the public good). Initially, Levi identified four main parallels between the rationale behind Bretton Woods and the needs of a climate regime that, 25 years later, still hold true: (a) the need for members to sacrifice economic sovereignty to provide a public good; (b) the need to facilitate capital investment and provide adjustment loans; (c) the need for equity and fairness; (d) the need for technical assistance and compliance monitoring.

Levi interpreted the Bretton Woods institutions (the International Monetary Fund – IMF, the World Bank and the WTO)[23] as built purposefully to deal with public non-excludable goods and to overcome any conflict between individual incentives and the public good. Following a logic in line with the MA proposal, Levi supported the idea of establishing a climate Bretton Woods. He underlined the need for establishing authorities similar to Bretton Woods institutions, as well as the need for designing a carbon accreditation system based on Tradable Emission Permits (TEPs) structured like the IMF reserves. This book translates Levi's findings into a governance proposal in today's context. Consequently, the relevance of CMO as investment asset is comparable to Levi's interpretation of TEP. Levi's description of the TEP system's possible monetary effects serves as a reference to understand the CMO system's possible effects on contemporary monetary systems.

Inspired by the functioning of the IMF's reserves, Levi envisaged TEPs as tradable international reserve assets convertible into local currencies. Levi suggested that climate alliance's members hold predetermined quantities of TEP reserve assets. By establishing specific quantities of TEP reserves, Levi equated the costs of policies to increase the earnings of this asset with the costs of policies to reduce spending on it, hence establishing TEP as the *carbon pricing* system for his climate Bretton Woods. Following Levi's view, the MA requires its members to hold a specific quantity of CMOs as reserve assets. Given that the MA's mitigation target can only be met using CMOs, virtually all the MA members shall be able to exchange

[23] Despite being founded only in 1995, the WTO was originally envisaged during the Bretton Woods discussions, hence interpreted by Levi as part of the Bretton Woods structure.

their currencies against the asset and all members that have been assigned a quota shall in fact posses a corresponding reserve of CMOs.

With all MA members valuing carbon equivalent to CMO, this latter one becomes their own currencies' reference currency, therefore allowing MA members to compare each other's economies by comparing their currencies' exchange rates against CMO (Mehrling 2013). The exchange rate of CMO against MA members' currencies, along with the CMO's double nature as price determiner in a market (when exchanged against other currencies) and policy instrument (when referring to its *commodity* - carbon) completes the definition of CMO as a *representative currency* for the MA.

Summarising the findings of Sects. 9.3 and 9.4, CMO holds a fixed exchange rate against a *commodity* (carbon), while MA Members' currencies maintain a flexible exchange rate against CMO and therefore also the *carbon commodity*. As a logical consequence, CMO serves as a unit of measure to compare the relative values of MA members' currencies. According to Yaacob and Ahmad (2014), this resembles a pseudo Bretton Woods commodity-based monetary system.

The Bretton Woods' commodity-based system, with currencies' values given by their exchange rate towards the U$D that held a fixed exchange rate towards gold,[24] was replaced by the fiat currency system in 1971. In the fiat currency system, still in use virtually everywhere in the world, the issuing governments back currency value with no specific link to any form of *commodity*. The fiat currency system allows (uncontrolled) multiplication of currency through fractional reserve banking and has been identified as the primary source of inflation and monetary instability (Frankman 2002; Yaacob and Ahmad 2014). Scholars have kept on calling for a return to a commodity-based system as the only feasible solution for the endemic problems associated with fiat currencies (Frankman 2002; D'Artista 2009; Yaacob and Ahmad 2014). Recently, this call has been increasingly linked to climate change policies (Hourcade et al. 2015; Pillay and Viñuales 2016; Aglietta and Espagne 2016; Campiglio 2016).

The MA proposal entails a *commodity*-based monetary system within its jurisdiction and establishes CMO as the *representative currency* for carbon. Similar to the Bretton Woods *commodity*-based system aimed at ensuring monetary stability and preserving the link between monetary issuance and the real economy (Graham 1944), the MA creates reserve assets that can provide stability in real value by monetising a primary *commodity* (carbon). This resembles the Bretton Woods reform proposed by Hart et al. (1964). In addition, the CMO *commodity*-based system has some unique features that further increase its significance as a financial and monetary tool. Firstly, by establishing a fixed banking quantity of CMOs (the MA NZC target), the MA reduces market volatility (Fell et al. 2012). Secondly, by allowing MA members and stakeholders to accumulate reserves of CMOs, the MA lowers overall mitigation costs (Fell et al. 2012). Thirdly, by being linked to an essentially public good, the CMO resembles a free liquidity type of currency, identified by Mehrling (2013) as the ideal type of currency that, bypasses any transaction cost

[24] See note 22.

typical of the currencies' issuance in fiat currency systems by being directly issued by the MAM to its users (Douthwaite 2012).

9.5 Conclusions: A Low Carbon Bretton Woods to Boost the Mitigation Alliance Benefits

Establishing CMO as a twenty-first century *representative currency* would represent a significant step towards a radical low carbon transformation of economy and society and may become a key benefit of the MA-led reform of the current climate regime and beyond. As described in Sect. 9.4, establishing CMO as a *representative currency* may harmonise MA members' currencies and, more in general, their economies, laying the foundations for a stable and durable alliance under a common goal in line with the PA's general objectives.

It is equally important to emphasise the step-by-step process of defining CMO as a *representative currency* for carbon, such as the definition of MA *carbon pricing* and the understanding of its importance as an instrument to attract private and public financial investments required to meet the MA target. With a NZC mitigation target by the end of the century and a CMO-based financial asset system established in support, it is now possible to envision a portfolio of indirect benefits supported by redirecting financial flows. Exploring such benefits entails speculation about the environmental, economic, social, and political effects driven by the MA, as well as investigation of actors that may benefit directly. Chapter 10 is entirely dedicated to this exploration and investigation.

References

Aglietta, M., & Espagne, E. (2016, April 10). *Climate and finance systemic risks, more than an analogy? The climate fragility hypothesis*. CEPII Working Paper.

Aglietta, M., Hourcade, J.-C., Jaeger, C., & Peressin Fabert, B. (2015). Financing transition in an adverse context: Climate finance beyond carbon finance. *International Environmental Agreements, 15*, 403–420.

Aldy, J. E., & Stavins, R. N. (2012). The promise and problems of pricing carbon: Theory and experience. *Journal of Environment and Development, 21*(2), 152–180.

Campiglio, E. (2016). Beyond carbon pricing: The role of banking and monetary policy in financing the transition to a low-carbon economy. *Ecological Economics, 121*, 220–230.

D'Artista, J. (2009). The evolving international monetary system. *Cambridge Journal of Economics, 33*, 633–652.

Dasgupta, D. (2016). 'De-carbonizing the world': Reviewing recent proposals on positive pricing of carbon and large-scale climate finance. In A. Sirkis (Ed.), *Moving the trillions: A debate on positive pricing of mitigation actions* (pp. 36–68). Rio de Janeiro: Brasil No Clima.

Douthwaite, R. (2012). Degrowth and the supply of money in an energy-scarce world. *Ecological Economics, 84*, 187–193.

Fell, H., MacKenzie, I. A., & Pizer, W. A. (2012). Prices versus quantities versus bankable quantities. *Resource and Energy Economics, 34*, 607–623.
Frankman, M. (2002). Beyond the tobin tax: Global democracy and a global currency. *Annals of the American Academy of Political and Social Science, 581*, 62–73.
Gollier, C., & Tirole, J. (2015). Negotiating effective institutions against climate change. *Economics of Energy and Environmental Policy, 4*(2), 5–27.
Graham, B. (1944). *World commodities and world currency*. New York: McGraw-Hill.
Hart, A. G., Kaldor, N., & Tinbergen, J. (1964). *The case for an international commodity reserve currency*. Geneva: UNCTAD.
Hourcade, J.-C., Shukla, P.-R., & Cassen, C. (2015). Climate policy architecture for the Cancun paradigm shift: Building on the lessons from history. *International Environmental Agreements, 15*, 353–367.
Hourcade, J.-C., Aglietta, M., Perissing Fabert, B., & Espagne, E. (2016). The low carbon transition between the animal spirits of finance and the fault lines of the world economy. In A. Sirkis (Ed.), *Moving the trillions: A debate on positive pricing of mitigation actions* (pp. 21–35). Rio de Janeiro: Brasil No Clima.
IEA. (2015). *World energy outlook 2015*. Paris: International Energy Agency (IEA).
IPCC. (2014). Climate change 2014: Synthesis report. Contribution of working groups I, II and III to the fifth assessment report of the Intergovernmental Panel on Climate Change [Core Writing Team, R.K. Pachauri, & L.A. Meyer (Eds.)]. Geneva: IPCC.
Jenkins, J. D. (2014). Political economy constraints on carbon pricing policies: What are the implications for economic efficiency, environmental efficacy, and climate policy design? *Energy Policy, 69*, 467–477.
Lanzi, E., Chateau, J., & Dellink, R. (2012). Alternative approaches for levelling carbon prices in a world with fragmented carbon markets. *Energy Economics, 34*, 240–250.
Levi, M. D. (1991). Bretton woods: Blueprint for a greenhouse gas agreement. *Ecological Economics, 4*, 253–267.
Marcu, A. (2016, January). Carbon market provisions in the Paris agreement (Article 6). *CEPS Special Report 128.*
Matthews, J. (2007). Seven steps to curb global warming. *Energy Policy, 35*, 4247–4259.
McCollum, D., Nagai, Y., Riahi, K., Marangoni, G., Calvin, K., Pietzcker, R., van Vliet, J., & van der Zwaan, B. (2013). Energy investments under climate policy: A comparison of global models. *LIMITS Special Issue*, FP7 LIMITS Project.
McKinnon, C. (2015). Climate justice in a carbon budget. *Climatic Change, 133*, 375–384.
Mehrling, P. (2013). Essential hybridity: A money view of FX. *Journal of Comparative Economics, 41*, 355–363.
Pagina22. (2013, October 24). The Rio climate challenge: Rio Clima, 2013. *Pagina 22* (Portuguese).
Pillay, K., & Viñuales, J. E. (2016). "Monetary" rules for a linked system of offset credits. *International Environmental Agreements: Politics, Law and Economics, 16*(6), 933–951.
Sirkis, A. (2016). Preface: The challenge of moving the trillions. In A. Sirkis (Ed.), *Moving the trillions: A debate on positive pricing of mitigation actions* (pp. 10–20). Rio de Janeiro: Brasil No Clima.
Stern, N. (2006, October 30). *Stern review on the economics of climate change*. Report to the Office of Climate Change, Government of United Kingdom, London, UK.
Stiglitz, J. E. (2015, July 11). *Global carbon pricing with cross-border tariffs needed to fight climate change*. Conference paper presented at Our Common Future under Climate Change. Paris.
Stua, M., Coulon, M., Nolden, C., & Sabljic, V. (2016). COP21 and beyond: Challenges for a fair agreement and the significance of the social and economic value of carbon mitigation actions and related positive carbon pricing. In A. Sirkis (Ed.), *Moving the trillions: A debate on positive pricing of mitigation actions* (pp. 113–143). Rio de Janeiro: Brasil No Clima.
Tietenberg, T. H. (2013). Reflections—Carbon pricing in practice. *Review of Environmental Economics and Policy, 7*(2), 313–329.
UNFCCC. (2015). *Adoption of the Paris Agreement*. Bonn: Climate Change Secretariat.

Weischer, L., Morgan, J., & Patel, M. (2012). Climate clubs: Can small groups of countries make a big difference in addressing climate change? *RECIEL, 21*(3), 177–192.

World Bank. (2016, October). *State and trends of carbon pricing*. World Bank Group Climate Change Report, Washington DC.

Yaacob, S. E., & Ahmad, S. (2014). Return to gold-based monetary system: Analysis based on gold price and inflation. *Asian Social Science, 10*(7), 18–28.

Zaman, P., Rock, N., Hedley, A., & Smokelin, J. (2016). *The Paris Agreement: Leading the pathway to a low carbon economy* (16 Platt's Energy Law Report 4, pp. 31–149). LexisNexis A.S. Pratt.

Chapter 10
Excludable Benefits Resulting from the Mitigation Alliance

Abstract The successful implementation of the Mitigation Alliance (MA) requires for its members to fully commit to the MA's aggregate, ambitious Net Zero Carbon (NZC) target. While it is unlikely for such a target to be based only upon the 'good will' of MA members, coercive measures to bind them to their assigned mitigation quotas (so–called 'sticks' within the climate club literature) would have no impact in a context where voluntary participation represents a key pillar for MA membership. Indeed the benefits offered by the recognition of the Certified Mitigation Outcome (CMO) as representative currency within the MA represent a significant stimulus for entities to converge towards the MA targets, yet this is not likely to suffice without additional, excludable co-benefits. As a matter of fact, the MA has been designed to guarantee a significant variety of co-benefits. While their effectiveness cannot be anticipated without verification of the effects of the MA, evidence resulting from previous climate governance experiments and the abundant literature discussing possible co-benefits linked to climate policies and strategies may well serve to outline the opportunities offered by the MA. This Chapter describes these opportunities and provides a first analysis of their actual effectiveness.

Keywords CMO value • Policies • Investments • Co-evolution • Excludable benefits

10.1 Introduction

An aggregate Net Zero Carbon (NZC) target distributed among Mitigation Alliance (MA) members over time necessitates corresponding quantities of carbon[1] emissions mitigation (or simply mitigation) outcomes. By recognising Certified Mitigation Outcomes (CMOs) as the only instrument to achieve the NZC target (and the corresponding assigned shares of burden or quotas[2] over time), the MA

[1] Carbon Dioxide (CO_2) is the most common Greenhouse Gas (GHG) and is often used as a reference unit to express levels of GHG emissions (expressed in CO_2 equivalent, or CO_2eq). In the context of this Chapter and, more in general, of this book, the term 'carbon', expressing CO_2eq, and the term GHG, are synonyms, hence carbon is used as single term to express GHG.

[2] Quota is used in this Chapter and more in general in this book as a synonym of 'assigned share of mitigation burden'.

M. Stua, *From the Paris Agreement to a Low-Carbon Bretton Woods*,
DOI 10.1007/978-3-319-54699-5_10

encourages policy innovation among its members for collecting and banking CMOs. This ensures that CMOs are assigned a value that attracts adequate finance and support for their production.

Within the MA, CMOs are both a tool to reach the MA mitigation target as well as a financial asset and a currency. The MA therefore aligns investor and policy maker commitment towards aggregate mitigation in line with the MA's NZC target despite often diverging interests.

Co-evolution of financial flows and policies within the MA provides exclusive benefits to MA members. Chapter 10 identifies the most significant exclusive benefits and explores their capacity as drivers for consolidating the MA. Chapter 10 also discusses their possible role as triggers for a transformation of societal and economic systems, which extends beyond realms of climate change. Three types of exclusive benefits ('private goods') are identified, based on the value distinction made by the Paris Agreement (PA). Once identified, a basic quantification of benefits is provided based on data from governance experiments.

Section 10.2 explores the co-evolutionary process triggered by the MA, providing also a comparison with similar efforts in recent years. Section 10.2 demonstrates how the foundation of the MA implies co-evolution and identifies in the application of MA rules the source of excludable benefits ('private goods') available to MA members and stakeholders.

Section 10.3 provides details on these benefits directly resulting from CMO lifecycle. It shows that the need for CMOs requires policies and investments to support the necessary resources for CMO production and supply. The necessary impact of CMO demand on such resources assigns CMOs with a relative value towards them. Section 10.3 analyses the value assigned to CMOs by adopting the discrimination scheme developed by Paragraph 108 of the Paris Agreement, which distinguishes between benefits of environmental, economic and social value.

Due to the novelty of the MA proposal, a quantitative analysis of the identified benefits' potential is not possible. Section 10.4 nevertheless attempts to offer a first clue on the possible magnitude of identified benefits by referring to data from past and present examples of climate experimental governance. Specifically, Sect. 10.4 refers to the vast literature on, inter alia, the impacts of tools such as the Clean Development Mechanism (CDM), Joint Implementation, Reducing Emissions from Deforestation and forest Degradation (REDD+), as well as of policy approaches such as Emission Trading Schemes (ETS) and carbon taxation, to support its findings.

In summarising the key findings of the Chapter, Sect. 10.5 attempts to provide a first understanding of the MA's transformational potential.

10.2 The Co-evolution of Policies and Investments Induced by the Mitigation Alliance

Since the very early debates on the effects of climate change, scholars and analyst have often attempted to demonstrate how early investments in mitigation actions pay off in terms of avoiding climate change (Nordhaus 1993; Stern 2006). Three main issues have arisen:

- Firstly, there is a problem regarding indemonstrability[3];
- Secondly, it faces a time dilemma[4];
- Thirdly, it leads to an impasse that Hardin (1968) defined as the 'tragedy of the commons'.[5]

In recent years, several climate scholars attempted to solve the 'trilemma' affecting the cost-based interpretation of mitigation by analysing potential co-benefits and converting them into excludable private goods for involved actors. Virtually the entire climate clubs literature[6] analyses the effectiveness of mitigation co-benefits as drivers for implementing mitigations. Similar analyses have also been conducted in the non-climate clubs literature (Bayer et al. 2013; Zhang and Shi 2014; Hourcade et al. 2015; Michaelowa 2015; Bodansky et al. 2016; Jakob et al. 2016).

In theory, the co-benefits approach demonstrates significant advantages compared to the cost-based approach. Yet, several barriers, such as the lack of ambition in parties' mitigation targets and the significant gaps between investments required for mitigation and investments provided, still limit the full implementation of a climate governance model based on the co-benefits approach. Co-benefits are often theoretical rather than practical. The MA aims to overcome these barriers by strengthening the impact of co-benefits for stakeholders involved in mitigation and beyond.

According to Das (2015, p. 24), 'to the extent (that) the club benefits succeed in generating climatic benefits, say greater climate mitigation, those would of course be public goods'. Successfully generating a public good, say greater climate mitigation, therefore necessitates private, excludable goods. In other words, by establishing a pathway leading towards its NZC mitigation target, the MA implicitly provides its members with the necessary and sustainable means to reach the target.

---

[3] Challenging the hypothesis of the need for early investments in practice would require continued unabated emission of carbon, which is likely to be irreversible given our current understanding of the environment and technology (IPCC 2014).

[4] The costs for mitigation would be incurred at present, while corresponding benefits are likely to be enjoyed by future generations.

[5] The public, non-excludable nature of mitigation allows 'free riding', as nothing is easier than benefitting from other's actions and avoiding paying the related mitigation costs. See also Chap. 3 of this book.

[6] See especially: Prakash and Potoski 2007; van't Veld and Kotchen 2011; Weischer et al. 2012; Stewart et al. 2013; Falkner 2016. See also Chap. 3 of this book.

As the MA can exist only once its members agree on their common mitigation target (and assigned shares of the mitigation burden, or quotas),[7] implementation of the MA requires its members to take action to meet their quotas. As CMOs issued by the Mitigation Alliance Mechanism (MAM) represent the only instrument recognised by the MA to assess members' progress towards meeting their quotas, MA members shall aim to accumulate a corresponding amount of CMOs.

MA members' demand for CMOs automatically assigns CMOs with a value equivalent to the cost of CMO accumulation. As discussed in Chap. 9 of this book, under these conditions, CMOs become an asset whose value can be calculated by "equating the costs of policies to increase the earnings of this asset with the costs of policies to reduce the spending of it" (Levi 1991, p. 261). Comparing the costs afforded by MA members to accumulate CMOs to fulfil their CMO quotas with the earnings obtained by CMO holders for supplying them to, the equation leads to a zero-sum result, where CMO holders gain as much as MA members pay.

While this zero-sum result may lead to the conclusion that any benefit is compensated by an equivalent cost, there are in fact several private, excludable goods for MA members and stakeholders resulting from the whole CMO lifecycle. Governed by a multi-level, hybrid governance model, the CMO lifecycle creates a certain and quantified demand, a corresponding supply through production, and multiple exchange approaches.

To sustain the CMO lifecycle, MA induced co-evolution of policies and investments converge towards supporting all sort of activities aimed at CMO lifecycle sustainability. Therefore potentially any stakeholder involved in the CMO lifecycle can become a recipient of multiple types of benefits resulting from the MA induced convergence.

10.3 Benefits Resulting from the Demand for CMO Production

CMOs do not exist *per se*. They only exist as by-products of actual, successful mitigation efforts. A demand for CMOs therefore implies for these efforts, such as successful CMO production and exchange to happen. To be effective, policies addressing the accumulation of CMOs must be coupled with policies favouring the necessary efforts to produce them. Similarly, revenues obtained by CMO exchange require previous investments in those efforts necessary for CMO production.

CMO production and exchange efforts require the use of variable proportions of resources including *inter alia* human, natural, technical, and financial resources. Hence, MA-induced demand for CMOs represents the main driver for corresponding demand of resources dedicated to CMO-related efforts. Acting as driver for resource demand, the original CMO demand provides these resources with *value*

[7] In the context of this Chapter and, more in general, of this book, the term *quota* and the term 'assigned shares of mitigation burden' are synonyms.

recognised among MA members and stakeholders. As the term *value* by itself may refer to various interpretations and as the MA intends to maintain alignment with the PA, CMO *value* can be categorised according to Paragraph 108 of the PA's recognition of *value* of voluntary mitigation outcomes: namely environmental, economic and social *value*.[8]

This distinction between CMO *value* categories attracts different interests, hence enhancing different co-evolutions of policies and investments to converge towards the resources favoured by the categories. Each of the categorised types of CMO *value* entails the distribution of specific benefits among stakeholders. This book distinguishes between two types of stakeholders included in the MA:

(a) direct stakeholders, providing resources required by the co-evolution process; and
(b) indirect stakeholders, benefitting from the combination of resource demand and their drivers.

Another significant categorisation relating MA induced co-evolutions of policies and investments concerns the possible relations between them. In principle, policies and investments relationships can be categorised into:

(a) coexisting (policies and investments move independently from each other);
(b) interacting (policies channel investments towards geographical, social, economic, environmental, and/or political priority areas); and
(c) corresponding (policies are based on direct investments).

Such categorisation is mainly based upon the different interests (or priorities) traditionally directing investments and policies in climate change and beyond.

Based on a possible simplistic synthesis it is here suggested that investments are associated with 'patience'. Private capital tends to be 'impatient' and associated investments motivated by the opportunity to gain in the short-term (Mazzucato 2015; Gaddy et al. 2017). Private investments are likely to be attracted by economically sustainable CMO production (or simply mitigation). Economic sustainability depends on a variety of factors, including subsidies allocated through policy intervention. Public sector capital tends to be more 'patient' (Mazzucato 2015). Economic sustainability might just be prospective, sometimes uncertain, when not limited to a supporting role. Subsidies fall into the category of 'patient' capital and are directly linked to policies.

Policies do not only focus on economic aspects and indeed do not only act through investments. Policies might have short or long-term targets and may encourage national, regional or local authorities to actively engage in CMOs production through industrial policies, welfare policies, energy security policies, environmental policies, and generally strategic policies. Policies might also target behaviours and only indirectly affect investments.

[8] Paragraph 108 of the PA recognises 'the social, economic and environmental value of voluntary mitigation actions and their co-benefits for adaptation, health and sustainable development' (Paragraph 108, Section IV of the PA).

Bearing in mind the above mentioned general features concerning policies and investments, while following the threefold categorisation of Paragraph 108 of the PA, the most significant benefits provided by climate-related policies and investments co-evolution to stakeholders include:

1. Environmental *value* benefits: (a) mitigation of carbon emissions[9]; (b) reduction of other pollutants[10]; (c) combat soil erosion and desertification[11]; (d) natural environment and species conservation.[12]
2. Economic *value* benefits: (a) improved investment conditions[13]; (b) stabilisation of the monetary system[14]; (c) shift of financial flows towards productive investments[15]; (d) enhancement of the cleantech innovation paradigm and increasing influence on other economic sectors.[16]
3. Social benefits include inter alia: (a) verifiable equity in mitigation[17]; (b) jobs and wealth creation[18]; (c) changes in individual and social behaviours; (d) diffused capacity building for low-carbon transformations; (e) health benefits[19]; (f) fiscal revenues increase.[20]

This list of benefits implies impacts on a significant variety of socio-technical regimes, including virtually all regimes traditionally related to climate change mitigation, adaptation, and resilience, as well as a potentially wide spectrum of regimes usually identified as not directly related to it. Inherently approximate estimations of

---

[9] Benefit inherent to CMOs.

[10] Depending on the types of mitigation, these benefits may include *inter alia*: air, water, and land pollutants; industrial residues; and general organic and inorganic waste.

[11] Benefits resulting from CMOs originated through forestry and REDD+ initiatives.

[12] Benefit as byproduct of environmental benefits (a), (b) and (c).

[13] Category including two types of benefits: (a) increased market liquidity and uncertainties and risk reduction due to the establishment of a steady, quantified demand for CMOs over time; and (b) lower transaction costs due to the adoption of harmonised standards and the markets linking processes.

[14] Benefit resulting from carbon pricing stability and through the institution of a representative currency.

[15] Implicit in the CMO nature, this benefit provides support to two industries: (a) manufacturing (i.e. asset owners; technology providers; component manufacturers); and (b) services (i.e. financial intermediaries; technical, legal, and accounting consultants; research, innovation, and development providers; capacity builders and education providers; MA bureaucracy and administration operators).

[16] These include multiple sub-sectors embedded in the three recognised macro-economic sectors: agriculture, industry and services.

[17] Benefit granted by the formula to distribute mitigation quotas among MA members (see Sect. 5.3 of Chap. 5 of this book).

[18] Benefits resulting from economic benefit 2(c).

[19] Benefits mainly resulting from environment-related benefits.

[20] Benefit resulting from economic benefit 2(c) and social benefit 3(a). It represents a significant window of opportunity for the implementation of climate-related policies and other policy priorities alike. Generally speaking, the benefit may also entail raising funds to support adaptation and resilience.

these benefits' impact on individuals and societies still represent a useful tool to provide a very first insight on the transformative capabilities of the MA.

10.4 Insights on the Potential Impact of the Mitigation Alliance

Providing reliable scenarios concerning the possible diffusion magnitude and characteristics of the MA impact on current regimes is not possible at present. Bearing in mind that an *a priori* analysis of a novelty that aims at transformational changes is simply unfeasible, and that the magnitude of the MA aggregate impact will mainly depend upon its capability to attract members, it is possible to make some assumptions on its potential impacts on individual MA members and stakeholders. Section 10.4 represents a non-exhaustive attempt to outline some of these assumptions.

To do so, Sect. 10.4 groups two sources of information: (a) empirical findings from the most significant climate governance experiments (including ETS, carbon taxes, offsetting mechanisms, and other policies); and (b) theoretical analyses from literature on the potentials of adequate climate governance strategies and actions. Representing both MA's references and objectives, to be reached and exceeded,[21] these two sources hence can be used as a first step to open a discussion on the MA's effective potential. By comparing the obtained results with the threefold categorisation of the benefits potentially resulting from climate-related actions,[22] Sect. 10.4 offers a first insight on the embryonic effectiveness of the MA.

As the MA requires effective mitigation, its most immediate impact will concern social and economic macro-sectors principally responsible for carbon emissions, including:

- Energy;
- Industrial process;
- Agriculture;
- Waste;
- Land-use change and forestry;
- Bunker fuels.

With its extensive number of key sub-sectors,[23] Energy represents by far the largest source of emission among the macro-sectors, with a share that has constantly remained close to three quarters of total carbon emissions over the last 20 years.[24] To meet its NZC target, Energy represents the core sector for CMO production.

---

[21] See also Chaps. 6, 7 and 8 of this book.

[22] Described in detail at the end of Sect. 10.3.

[23] Including: electricity and heat; manufacturing and construction; transportation; other fuel combustion; and general fugitive emissions.

[24] Source: World Resource Institute—WRI (http://cait.wri.org/).

Hence evidence from already implemented climate policies in the Energy sector can provide an outlook on the potential of MA climate governance.

Where applied, experimental climate-related governance produced significant effects on Energy and its sub-sectors. For instance, a significant amount of information concerning the potential of carbon mitigation mechanisms can be identified by analysing the data referring to the 3642 CDM projects hosted in China from 2005 to 2012.[25] In their vast majority concentrated on activities related to electricity production, Chinese CDM projects added more than 180GW of electricity capacity from renewable, clean, or efficient energy sources. The CDM became a core tool for the development of new energy industries in China and leveraged direct investments for more than U$D 200 billions on aggregate (Stua 2013).

Extending the breadth of macro-sectors and geographic areas, climate literature offers insights into the capability for tools like CDM, REDD+, ETS, voluntary carbon markets, and carbon taxation, as well as the effectiveness of climate policy experimentation, to provide effective co-benefits to a multiplicity of stakeholders by:

(a) Shifting financial flows towards low-carbon, productive sectors (Grubb et al. 2011; Gujba et al. 2012; Mez 2012; Zavodov 2012: Bayer et al. 2013; Frankhauser 2013; Springmann 2013; Gawel et al. 2014; Lim and Lam 2014; Mazzucato 2015; Buchsbaum 2016);
(b) Creating jobs and combating poverty, including energy poverty (Rive and Rübbleke 2010; Shreshta and Shakya 2012; Wang et al. 2013);
(c) Reducing pollutants and increasing human health and safety (Jia et al. 2013; Uddin et al. 2015; Murata et al. 2016);
(d) Stimulating civil society participation and stakeholders' engagement (Bedsworth and Hanak 2013; Mustalahti and Rakotonarivo 2014; Nolden 2015; Bernauer et al. 2016; Kuchler and Lövbrand 2016);
(e) Creating synergies between social and environmental benefits (Khatun 2013; La Rovere et al. 2013; Olsson and Ouattara 2013; McGregor et al. 2015; Poudyal et al. 2016);
(f) Driving technological innovation and technology transfer (de Coninck et al. 2007; Van der Gaast et al. 2009; Rogge et al. 2011; Lim and Kim 2012; Lema and Lema 2013; Gandenberger et al. 2015; Potdar et al. 2016).

Despite the evidence of such benefits, past and current experiments in climate governance have not yet unleashed the full potential of mitigation in terms of co-benefits. As already discussed in several Chapters of this book, current climate governance is frequently paralysed by excessive transaction costs, fragmentation, weak equity and, most of all, lack of ambition, which altogether challenge the sustainability of mitigation actions, policies and strategies (Subbarao and Lloyd 2011;

[25] Data concerning Chinese CDM projects were collected by the main author of this book through the analysis of the corresponding 3642 Project Designed Documents (PDDs) submitted by the project owners to the CDM Executive Board (see also Sect. 6.1 of Chap. 6 of this book). PDDs of all CDM projects are available at: https://cdm.unfccc.int/.

Markandya 2011; Castro 2012; Mattoo and Subramanian 2012; Hayashi and Michaelowa 2013; Tatrallyay and Stadelmann 2013; Ervine 2014; Trotter et al. 2015; Dirix et al. 2016; Zhang and Sun 2016).

The multiple innovations generated by the adoption of an ambitious mitigation target, the establishment of a single and common system to certify mitigation outcomes, and the guarantee of equal rights in managing CMOs for all the MA stakeholders may provide the MA with the capacity to enhance carbon mitigation in line with PA targets, hence leading to a co-evolution of co-benefits. Widely theorised by climate literature, these co-benefits, which have not been achieved through existing climate governance regimes to date, include:

(a) Decoupling economic growth from emissions (Brinkley 2014; Zhang and Shi 2014);
(b) Linking mitigation to adaptation and resilience[26] (Goklany 2007; Felgenhauer and Webster 2013; Casado-Asensio and Steurer 2014; Pauw 2015);
(c) Promoting co-evolution of economic growth, innovation, and welfare (Frankhauser et al. 2008; Acemoglu et al. 2012; Gans 2012; Hallegatte et al. 2012; Lund and Hvelplund 2012);
(d) Widening commodification between environmental, economic, and social policies (Kok and de Coninck 2007; Warnecke et al. 2014; Pittman et al. 2015);
(e) Transforming current energy systems (Vad Mathiesen et al. 2011; Cooper 2016; Stua et al. 2016);
(f) Supporting power shift currently in action worldwide (Timmons Roberts 2011; Hurrell and Sengupta 2012; Kim 2012; Hochstetler and Milkoreit 2014).

While the MA's impact on carbon mitigation and monetary innovation described elsewhere in this book has already provided evidence on the overall ambitions of the MA proposal, this condensed summary of co-benefits resulting from MA implementation represents a reliable description of its strength in guaranteeing cohesion and collaboration among MA members. The varied list of co-benefits for MA members and stakeholders indeed represents the most practical translation of private, excludable goods as stressed by climate clubs literature, when defining the ideal architecture of an effective climate governance structure.[27] Thanks to the elements discussed in Chaps. 9 and 10 of this book it is possible to affirm that the MA entails all the four attributes (speed, ambition, participation, and equity) required for an effective, radical transformation of the climate-related regime (Biermann et al. 2009; Weischer et al. 2012).

[26] As often recommended, if not imposed, by the PA (see Articles 4, 5, 6, 7, 9, 10, 11 and 13 of the PA).

[27] See also Chap. 3 of this book.

10.5 Conclusions: Transformational Potential of the Mitigation Alliance

Summarising the findings of this Chapter, as well as of the whole analysis developed in this book, it appears manifest that, if fully and correctly implemented, the MA may effectively merge and enhance all elements required for a deep decarbonisation and low carbon transformation of the current social and economic regimes.

A full implementation of the MA may significantly drive radical environmental change, radical economic innovation, and radical social transformation. Nevertheless, the MA entails also a significant set of risks, uncertainties and limits, embedded within its own ambition. Chapter 11, the final Chapter of this book, will discuss such barriers, which may significantly challenge the future development of a MA-oriented governance model.

References

Acemoglu, D., Aghion, P., Bursztyn, L., & Hemous, D. (2012). The environment and directed technical change. *American Economic Review, 102*, 131–166.

Bayer, P., Marcoux, C., & Urpelainen, J. (2013). Leveraging private capital for climate mitigation: Evidence from the clean development mechanism. *Ecological Economics, 96*, 14–24.

Bedsworth, L. W., & Hanak, E. (2013). Climate policy at the local level: Insights from California. *Global Environmental Change, 23*, 664–677.

Bernauer, T., Gampfer, R., Meng, T., & Su, Y.-S. (2016). Could more civil society involvement increase public support for climate policy-making? Evidence from a survey experiment in China. *Global Environmental Change, 40*, 1–12.

Biermann, F., Pattberg, P., van Asselt, H., & Zelli, F. (2009). The fragmentation of global governance architectures: A framework for analysis. *Global Environmental Politics, 9*(1), 14–25.

Bodansky, D. M., Hoedl, S. A., Metcalf, G. E., & Stavins, R. N. (2016). Facilitating linkage of climate policies through the Paris outcome. *Climate Policy, 16*(8), 956–972.

Brinkley, C. (2014). Decoupled: Successful planning policies in countries that have reduced per capita greenhouse gas emissions with continued economic growth. *Environment and Planning C: Government and Policy, 32*, 1083–1099.

Buchsbaum, L. (2016). Germany's Energiewende at a new turning point. *Power, 160*(3), 18–27.

Casado-Asensio, J., & Steurer, R. (2014). Integrated strategies on sustainable development, climate change mitigation and adaptation in Western Europe: Communication rather than coordination. *Journal of Public Policy, 34*(3), 437–473.

Castro, P. (2012). Does the CDM discourage emission reduction targets in advanced developing countries? *Climate Policy, 12*, 198–218.

Cooper, M. (2016). Renewable and distributed resources in a post-Paris low carbon future: The key role and political economy of sustainable electricity. *Energy Research & Social Science, 19*, 66–93.

Das, K. (2015). Climate clubs: Carrots, sticks and more. *Economic & Political Weekly, 50*(34), 24–27.

De Coninck, H., Haake, F., & van der Linden, N. (2007). Technology transfer in the clean development mechanism. *Climate Policy, 7*, 444–456.

Dirix, J., Peeters, W., & Sterckx, S. (2016). Is the clean development mechanism delivering benefits to the poorest communities in the developing world? A critical evaluation and proposals for reform. *Environment, Development and Sustainability, 18*(3), 839–855.

Ervine, K. (2014). Diminishing returns: Carbon market crisis and the future of market- dependent climate change finance. *New Political Economy, 19*(5), 723–747.

Falkner, R. (2016). A minilateral solution for global climate change? On bargaining efficiency, club benefits and international legitimacy. *Perspectives on Politics, 14*(1), 87–101.

Felgenhauer, T., & Webster, M. (2013). Multiple adaptation types with mitigation: A framework for policy analysis. *Global Environmental Change, 23*, 1556–1565.

Frankhauser, S. (2013). A practitioner's guide to a low-carbon economy: Lessons from the UK. *Climate Policy, 13*(3), 345–362.

Frankhauser, S., Sehlleier, F., & Stern, N. (2008). Climate change, innovation and jobs. *Climate Policy, 8*, 421–429.

Gaddy, B., Sivaram, V., Jones, T., & Wayman, L. (2017). Venture capital and cleantech: The wrong model for energy innovation. *Energy Policy, 102*, 385–395.

Gandenberger, C., Bodenheimer, M., Schleich, J., Orzanna, R., & Macht, L. (2015). Factors driving international technology transfer: Empirical insights from a CDM project survey. *Climate Policy*. doi:10.1080/14693062.2015.1069176.

Gans, J. (2012). Innovation and climate change policy. *American Economic Journal: Economic Policy, 4*(4), 125–145.

Gawel, E., Strunz, S., & Lehmann, P. (2014). A public choice view on the climate and energy policy mix in the EU — How do the emissions trading scheme and support for renewable energies interact? *Energy Policy, 64*, 175–182.

Goklany, I. M. (2007). Integrated strategies to reduce vulnerability and advance adaptation, mitigation, and sustainable development. *Mitigation and Adaptation Strategies for Global Change, 12*(5), 755–786.

Grubb, M., Laing, T., Counsell, T., & Willan, C. (2011). Global carbon mechanisms: Lessons and implications. *Climatic Change, 104*, 539–573.

Gujba, H., Thorne, S., Mulugetta, Y., Rai, K., & Sokona, Y. (2012). Financing low carbon energy access in Africa. *Energy Policy, 47*, 71–78.

Hallegatte, S., Heal, G., Fay, M., & Treguer, D. (2012). *From growth to green growth: A framework* (World Bank policy research working paper, 5872), Washington, DC.

Hardin, G. (1968). The tragedy of the commons. *Science, 162*(3859), 1243–1248.

Hayashi, D., & Michaelowa, A. (2013). Standardization of baseline and additionality determination under the CDM. *Climate Policy, 13*(2), 191–209.

Hochstetler, K., & Milkoreit, M. (2014). Emerging powers in the climate negotiations: Shifting identity conceptions. *Political Research Quarterly, 67*(1), 224–235.

Hourcade, J.-C., Shukla, P.-R., & Cassen, C. (2015). Climate policy architecture for the Cancun paradigm shift: Building on the lessons from history. *International Environmental Agreements, 15*, 353–367.

Hurrell, A., & Sengupta, S. (2012). Emerging powers, north–south relations and global climate politics. *International Affairs, 88*(3), 463–484.

IPCC. (2014). Climate Change 2014: Synthesis report. Contribution of working groups I, II and III to the *Fifth Assessment Report of the Intergovernmental Panel on Climate Change* [Core Writing Team, R. K. Pachauri, & L. A. Meyer (Eds.)]. IPCC, Geneva, Switzerland.

Jakob, M., Chen, C., Fuss, S., Marxen, A., Rao, N. D., & Edenhofer, O. (2016). Carbon pricing revenues could close infrastructure access gaps. *World Development, 84*, 254–265.

Jia, R., Guo, X., & Marinova, D. (2013). The role of the clean development mechanism in achieving China's goal of a resource-efficient and environmentally friendly society. *Environment, Development and Sustainability, 15*(1), 133–148.

Khatun, K. (2013). Integrating national forestry initiatives in India with international climate change policy. *Climate Policy, 13*(3), 384–402.

Kim, S. (2012). Japan's 'common but differentiated' approach to sustainable development and climate change in Africa. *Japanese Studies, 32*(2), 255–274.

Kok, M. T. J., & de Coninck, H. C. (2007). Widening the scope of policies to address climate change: Directions for mainstreaming. *Environmental Science and Policy, 10*, 587–599.

Kuchler, M., & Lövbrand, E. (2016). Simulative governance: On the collaborative narrative of civil society participation in the CDM stakeholder framework. *Environmental Politics, 25*(3), 434–453.

La Rovere, E. L., Dubeux, C. B., Pereira Jr., A. O., & Wills, W. (2013). Brazil beyond 2020: From deforestation to the energy challenge. *Climate Policy, 13*(1), 70–86.

Lema, A., & Lema, R. (2013). Technology transfer in the clean development mechanism: Insights from wind power. *Global Environmental Change, 23*, 301–313.

Levi, M. D. (1991). Bretton woods: Blueprint for a greenhouse gas agreement. *Ecological Economics 4*, 253–267.

Lim, J.-S., & Kim, Y.-G. (2012). Combining carbon tax and R&D subsidy for climate change mitigation. *Energy Economics, 34*, 496–502.

Lim, X.-L., & Lam, W.-H. (2014). Review on clean development mechanism (CDM) implementation in Malaysia. *Renewable and Sustainable Energy Reviews, 29*, 276–285.

Lund, H., & Hvelplund, F. (2012). The economic crisis and sustainable development: The design of job creation strategies by use of concrete institutional economics. *Energy, 43*, 192–200.

Markandya, A. (2011). Equity and distributional implications of climate change. *World Development, 39*(6), 1051–1060.

Mattoo, A., & Subramanian, A. (2012). Equity in climate change: An analytical review. *World Development, 40*(6), 1083–1097.

Mazzucato, M. (2015). Innovation, the state and patient capital. *The Political Quarterly, 86*(1), 98–118.

McGregor, A., Challies, E., Howson, P., Astuti, R., Dixons, R., Haalboom, B., Gavin, M., Tacconi, L., & Afiff, S. (2015). Beyond carbon, more than forest? REDD+ governmentality in Indonesia. *Environment and Planning, 47*, 138–155.

Mez, L. (2012). Germany's merger of energy and climate change policy. *Bulletin of the Atomic Scientists, 68*(6), 22–29.

Michaelowa, A. (2015). Opportunities for and alternatives to global climate regimes post-Kyoto. *Annual Review of Environment and Resources, 40*, 395–417.

Murata, A., Liang, J., Eto, R., Tokimatsu, K., Okajima, K., & Uchiyama, Y. (2016). Environmental co-benefits of the promotion of renewable power generation in China and India through clean development mechanisms. *Renewable Energy, 87*, 120–129.

Mustalahti, I., & Rakotonarivo, O. S. (2014). REDD+ and empowered deliberative democracy: Learning from Tanzania. *World Development, 59*, 199–211.

Nolden, C. (2015). *Performance and impact of the Feed-in Tariff scheme, Review of evidence*. Final report. Department of Energy and Climate Change (DECC), London.

Nordhaus, W. D. (1993). Reflection of economics of climate change. *Journal of Economic Perspectives, 7*(4), 11–25.

Olsson, E. G. A., & Ouattara, S. (2013). Opportunities and challenges to capturing the multiple potential benefits of REDD+ in a traditional transnational savanna-woodland region in West Africa. *Ambio, 42*, 309–319.

Pauw, W. P. (2015). Not a panacea: Private-sector engagement in adaptation and adaptation finance in developing countries. *Climate Policy, 15*(5), 583–603.

Pittman, J., Armitage, D., Alexander, S., Campbell, D., & Alleyne, M. (2015). Governance fit for climate change in a Caribbean coastal-marine context. *Marine Policy, 51*, 486–498.

Potdar, A., Singh, A., Unnnikrishnan, S., Naik, N., Naik, M., & Nimkar, I. (2016). *Process Safety and Environmental Protection, 101*, 160–169.

Poudyal, M., Ramamonjisoa, B. S., Hockley, N., Rakotonarivo, O. S., Gibbons, J. M., Mandimbiniaina, R., Rasoamanana, A., & Jones, J. P. G. (2016). Can REDD+ social safeguards reach the 'right' people? Lessons from Madagascar. *Global Environmental Change, 37*, 31–42.

Prakash, A., & Potoski, M. (2007). Collective action through voluntary environmental programs: A club theory perspective. *Policy Studies Journal, 35*(4), 773–792.
Rive, N., & Rübbleke, D. T. G. (2010). International environmental policy and poverty alleviation. *Review of World Economics, 146*, 515–543.
Rogge, K. S., Schneider, M., & Hoffmann, V. H. (2011). The innovation impact of the EU emission trading system — Findings of company case studies in the German power sector. *Ecological Economics, 70*, 513–523.
Shreshta, R. M., & Shakya, S. R. (2012). Benefits of low carbon development in a developing country: Case of Nepal. *Energy Economics, 34*, 503–512.
Springmann, M. (2013). Carbon tariffs for financing clean development. *Climate Policy, 13*, 20–42.
Stern, N. (2006, October 30). *Stern review on the economics of climate change*. Report to the Office of Climate Change, Government of United Kingdom, London, UK.
Stewart, R., Oppenheimer, M., & Rudyk, B. (2013). A new strategy for global climate protection. *Climatic Change, 120*, 1–12.
Stua, M. (2013). Evidence of the clean development mechanism impact on the Chinese electric power system's low-carbon transition. *Energy Policy, 62*, 1309–1319.
Stua, M., Nolden, C., & Costa, A. (2016). What if… Discussing the effects of the Paris Agreement on the energy markets. *Proceedings of the 23th World Energy Congress 2016*. Istanbul.
Subbarao, S., & Lloyd, B. (2011). Can the clean development mechanism (CDM) deliver? *Energy Policy, 39*, 1600–1611.
Tatrallyay, N., & Stadelmann, M. (2013). Climate change mitigation and international finance: The effectiveness of the clean development mechanism and the global environment Facility in India and Brazil. *Mitigation and Adaptation Strategies for Global Change, 18*(7), 903–919.
Timmons Roberts, J. (2011). Multipolarity and the new world (dis)order: US hegemonic decline and the fragmentation of the global climate regime. *Global Environmental Change, 21*, 776–784.
Trotter, I. M., da Cunha, D. A., & Féres, J. G. (2015). The relationships between CDM project characteristics and CER market prices. *Ecological Economics, 119*, 158–167.
Uddin, N., Blommerde, M., Taplin, R., & Laurence, D. (2015). Sustainable development outcomes of coal mine methane clean development mechanism projects in China. *Renewable and Sustainable Energy Reviews, 45*, 1–9.
Vad Mathiesen, B., Lund, H., & Karlsson, K. (2011). 100% renewable energy systems, climate mitigation and economic growth. *Applied Energy, 88*, 488–501.
Van der Gaast, W., Begg, K., & Flamos, A. (2009). Promoting sustainable energy technology transfers to developing countries through the CDM. *Applied Energy, 86*, 230–236.
Van't Veld, K., & Kotchen, M. (2011). Green clubs. *Journal of Environmental Economics and Management, 62*, 309–322.
Wang, C., Zhang, W., Cai, W., & Xie, X. (2013). Employment impacts of CDM projects in China's power sector. *Energy Policy, 59*, 481–491.
Warnecke, C., Wartmann, S., Höhne, N., & Bolk, K. (2014). Beyond pure offsetting: Assessing options to generate net-mitigation-effects in carbon market mechanisms. *Energy Policy, 68*, 413–422.
Weischer, L., Morgan, J., & Patel, M. (2012). Climate Clubs: Can small groups of countries make a big difference in addressing climate change? *RECIEL, 21*(3), 177–192.
Zavodov, K. (2012). Renewable energy investment and the clean development mechanism. *Energy Policy, 40*, 81–89.
Zhang, Y., & Shi, H.-L. (2014). From burden-sharing to opportunity-sharing: Unlocking the climate negotiations. *Climate Policy, 14*(1), 63–81.
Zhang, Y.-J., & Sun, Y.-F. (2016). The dynamic volatility spillover between European carbon trading market and fossil energy market. *Journal of Cleaner Production, 112*, 2654–2663.

Chapter 11
Uncertainties, Risks and Limits

Abstract To conclude this long and comprehensive description of the MA structure, functioning and governance model, as well as of its potential effects and benefits, this chapter attempts to describe those elements that most likely can and will represent significant obstacles for the MA's full implementation. Three categories of issues are here identified: uncertainties, risks, and limits. Uncertainties represent issues that can and will be solved only once the MA itself will be implemented. Risks represent disturbance elements that will affect the full and correct implementation of the MA. Limits represent barriers that the MA either cannot anticipate or effectively solve throughout its existence. While uncertainties and risks can be managed within the overall framework of the MA, limits represent a higher level threat to the functioning of the MA, which will require efforts beyond the MA domain to address successfully.

Keywords Uncertainties • Risks • Limits • Critical mass • Relevant members • Tragedy of commons • Common good • MA timeframe dilemma

11.1 Introduction

To conclude this Mitigation Alliance (MA) climate governance model it is necessary to investigate which potential issues may eventually affect its design and successful implementation.

To do so, this chapter identifies three separate categories of concerns, dedicating a section to each of them. Section 11.2 identifies the main uncertainties that affect the MA proposal in its current status. The proposal outlined in this book in fact includes multiple elements concerning the MA structure and functioning that cannot be solved in a theoretical and *a priori* perspective. Section 11.2 attempts to frame uncertainties regarding the MA proposal within a more empirical, policy oriented approach.

Analysing the MA main risks, Sect. 11.3 discusses issues that most likely will result from the MA implementation. In their greatest part already introduced in other Chapters of this book, such risks represent potential negative effects resulting from the MA.

M. Stua, *From the Paris Agreement to a Low-Carbon Bretton Woods*,
DOI 10.1007/978-3-319-54699-5_11

In describing the MA limits, Sect. 11.4 aims to identify the most significant issues falling beyond the MA domain.

11.2 Uncertainties Affecting the Mitigation Alliance Proposal

Uncertainty is defined as a situation in which something is not known, or what is known about something is certain. The scope and breath characterising the MA proposal, coupled with the inclusion of a consistent set of novelties and innovations in the field of climate change policy and governance, as well as with the MA long-term perspective imply the existence of several uncertainties that, in their vast majority, will remain unsolved until the MA will be effectively adopted and implemented. This section does not aim at providing a full list of such uncertainties, but rather focuses on two specific examples that potentially resemble the greatest challenges for the full and correct implementation of the MA: (a) critical mass of the MA; (b) enforcement measures.

The critical mass issue represents by far the biggest uncertainty threatening MA implementation. Voluntary participation in the MA represents one of the core pillars for the MA proposal. Voluntary participation embodies a pre-condition for the MA implementation: only once potential members freely and formally agreed their participation[1] can the MA be instituted and implemented. At the same time, the MA's effectiveness will heavily depend upon its capability to attract varied wide range of members.

While describing the ideal structure for a transformational club, Weischer et al. (2012, p. 179) state that this club should be able to "provide a proof of concept for low-emissions development, accelerate technology development and diffusion, create momentum and catalyze other initiatives." In identifying the ideal subjects to be included as members of the transformational club, Weischer et al. (2012, p. 179) introduce the concept of 'relevant' in the context of providing such proof of concept as the transformational club "needs to involve a critical mass of 'relevant' countries." Aside from relevance in terms of emissions, Weischer et al. (2012, p. 179) differentiate between three different types of 'relevance' for countries' representation inside the club: (a) political or strategic relevance; (b) economic relevance; (c) symbolic relevance.

To meet its primary ambition, represented by the capability to significantly impact the global levels of carbon emission, the MA must be able to attract a critical mass of members where all three types of 'relevance' identified by Weischer et al. (2012, p. 179) are fully represented. Such representation would help the MA to fully develop its potentials, as well as encourage MA participation over time.

Uncertainty linked to the capability of the MA to reach its *critical mass of relevant members* cannot at this moment be fully clarified through a proof of evidence

---

[1] As stated in Paragraph 1 of Article 6 of the Paris Agreement (A6PA). See also Sect. 8.2 of Chap. 8 of this book.

corresponding to the MA's implementation. Yet elements discussed inside this book allow some considerations regarding the rationale supporting the idea of a MA capable of engaging a significant and relevant set of members over time. The ambitious Net Zero Carbon (NZC) target and the necessary review of the sovereignty concept in climate change governance represent the two greatest barriers for a full implementation of the MA as a catalyser of transformational activities.

The target may represent an excessive burden in the eyes of potential members as demonstrated by past experience in climate governance.[2] Yet the momentum established in Paris apparently demonstrated a renewed willingness to move towards strong and effective *global climate action*. In addition, to counterbalance the potential negative impact of an ambitious target over the willingness to participate, the MA guarantees an equitable system of distribution of the target[3] and a multitude of opportunities to meet it.[4]

The sovereignty limitation resulting from the setting of a top-down-style aggregate target and the foundation of international authorities partly responsible for MA governance may equally represent a disincentive for MA membership. Regarding this second element of uncertainty the Paris Agreement does not provide explicit support. In fact, the PA may oppose top-down climate governance by favouring an apparently more bottom-up governance model based on Nationally Determined Contributions (NDCs). As a response to the worries concerning sovereignty, the MA proposes a profoundly new, multi-layered, hybrid model for climate governance, where MA authorities, members, and stakeholders are equally involved in the general governance process and where decisions, strategies, policies, and actions can be taken at any level of decision-making.[5]

Yet there is little doubt that the capability for the MA to attract members and catalyse attention at the global level will mainly depend on the attractiveness of its theorised public and private benefits.[6] Only confidence regarding the effective ability of the MA to provide such benefits may in fact encourage participation in line with the MA's ambitions.

The second relevant uncertainty surrounding the MA concerns its enforcement structure. While describing the Mitigation Alliance Target Fund (MATF),[7] Sect. 8.6 of Chap. 8 of this book defines a variety of enforcement measures for guaranteeing the MA's target achievement,[8] without expressly discriminating among them. The

[2] For instance the withdrawal of the United States of America from the Kyoto Protocol (KP) in 2001.

[3] See Sect. 5.4 of Chap. 5 of this book.

[4] See Chaps. 6 and 7 of this book.

[5] See Chap. 8 of this book.

[6] See Chaps. 9 and 10 of this book.

[7] The MATF represents the authority in charge for the MA mitigation target.

[8] The proposed enforcement options include: (a) monetary fee; (b) rating; (c) exclusion; (d) self-empowerment through the recognition of excludable co-benefits. For further details see Sect. 8.6 of Chap. 8 of this book.

MA proposal leaves the final decision concerning the optimal enforcement solution to policy makers who will be directly involved in the MA implementation.

Despite not being exclusionary, hence possibly overlapping, the identified enforcement options are significantly different from one another, both in terms of potential impact on the MA members and in terms of possible effects on the MA implementation as a whole. The options described in Sect. 8.6 of Chap. 8 of this book in fact include different degrees of penalty intensity, hence possibly influencing both the MA's effectiveness and its capability to function as a critical mass catalyser. Excessively tight enforcement measures may represent an aggregation disincentive, reducing the MA attractiveness, while loose measures may be unable to grant the MA sufficient capability to meet its ambitions.

As the final decision over MA enforcement measures is left to policy making decisions, there is no chance to *a priori* foresee which of the available options will be finally adopted nor the effects of such adoption. Nevertheless, similarly to the discussion concerning critical mass creation, the capability of MA members to recognise the benefits resulting from the MA proposal may significantly impact the establishment of a reliable enforcement system. The greater the perception of benefits, the more likely it will be for members to follow a self-empowerment approach, where meeting assigned mitigation will be perceived as the 'fair fee' for obtaining benefits.

11.3 Risks Resulting from the Mitigation Alliance Proposal

Compared to uncertainties, which cannot be fully identified and solved before the implementation of the MA, its most significant risks can be easily identified and eventually targeted with appropriate counteractions even before the institution of the MA. Section 11.3 identifies four main risks related to the MA:

1. Coexistence with international legislation;
2. Losers;
3. Educational issues;
4. The tragedy of the commons.

The first risk has been widely discussed in several sections of this book, especially Sects. 8.2 (the relationship between the MA and international climate legislation) and 8.3 (the relationship with other categories of international legislation) of Chap. 8. To many extents, these sections address some of the biggest risks related to the potential conflicts within different law regimes. Nevertheless, the ambition, novelty, and potential breadth of the MA may lead to conflicts over time. Additional studies and research that extend the understanding of the possible relationships between the MA governance model and international legislation represents the only effective way to reduce and prevent risks of conflicts regarding international legislation.

Implementation of the MA is likely to create 'losers', in terms of stakeholders whose gains from the MA do not exceed costs. Losers may be identified in specific socio/economic sectors threatened by the ambitions of the MA (i.e. the petrochemical industry and more in general the fossil fuels industry) and in geographical areas heavily relying upon such sectors. While there is no solution capable of safeguarding the sectors affected by the transformation envisaged by the MA as a result of their primary responsibility for the climate dilemma in the first place, specific strategies and policies aimed at supporting the 'losers' within these sectors can be envisaged and implemented both at MA members level and on aggregate.

Reconversion programmes, training, and capacity building strategies and plans for geographical equity represent the measures that are most likely to support a sustainable transition of MA losers. As stated above, similar initiatives can be implemented at different levels: individual members and stakeholders may adopt measures in this sense, while an aggregate MA programme supporting 'losers', eventually founded through shares of proceeds or similar approaches,[9] may further promote equity within the system.

Educational risk represents one of the biggest threats for the social sustainability of the MA. An effective implementation of the MA may lead to a radical transformation of the socio-economic systems of MA members in a relatively short-term, hence requiring significant change also in the competences, capabilities, skills, behaviours, and habits of the involved labour force and, more in general, of the whole population of MA members. Such a radical shift requires the establishment of adequate educational and training systems at any possible level, from the structures devoted to the preparation of an appropriate skilled workforce to basic education in schools. Such a significant educational task requires the simultaneous and effective establishment of a variety of efforts: coordinated education programmes, significant investments to 'train the trainers', and to provide the required software and hardware, political coordination, and planning.

All MA members and other stakeholders will need to set up and implement their educational programmes capable of integrating short-, medium- and long-term perspectives. It is unlikely for such multiple implementation systems to develop homogeneously both within and between the members and stakeholders, hence possibly leading to discrepancies and disparities in the MA system, which may lead to instability and destabilisation in the long run. Possible solutions may lie in integrated educational approaches at MA level, aimed at guaranteeing an efficient, even, and equitable development of the capabilities and skills required for MA implementation among all its members.

The 'tragedy of the commons' is an economic theory of a situation within a shared-resource system where individual users acting independently according to their own self-interest behave contrary to the *common good* of all users by depleting that resource through their collective action (Hardin 1968). A similar risk is inherently and naturally present in the MA, which is founded on the concept of converging

[9] For a detailed description of approaches that may be eventually adopted to support MA 'losers' see also Chaps. 7 and 8 of this book.

towards a *common good*. The rationale of this book is to solve the dilemma of the 'tragedy of commons' by encouraging individuals to act for the *common good* while pursuing their own, private interests. The benefits envisaged in Chaps. 9 and 10 of this book shall eventually stimulate individuals to act according to the MA goal (towards a common good), while still maintaining focus on their own 'goods'. Indeed this represents the biggest challenge for the MA and the risk for a failure.

Resulting from either the MA's inability to 'transmit the message' to its members and stakeholders, or by their misunderstanding of this message, the 'tragedy of commons' can and will be contained and overcome only if and when the MA's impact will be undeniable and its promised benefits will be fully and equally provided.

11.4 Limits of the Mitigation Alliance Proposal

As any other attempt to solve a set of current issues, and despite its ambitious perspective, the MA includes inherent limits that cannot be overcome within its own context. Unlike the uncertainties and risks described above, the MA limits can be identified but cannot be solved. At the current stage it is possible to envisage two significant limits to the MA proposal:

1. The lack of actions facing climate change related loss and damage;
2. The MA timeframe dilemma.

The MA proposal is by its nature focused on climate issues that relate to the mitigation concept. In other words, it aims to mitigate emissions, hence reducing the impact of climate change on human beings and nature. Nevertheless, some aspects of the MA proposal include elements related to other climate domains, such as adaptation and resilience. On the one side, the MA system will directly support initiatives combining mitigation, adaptation, and resilience. On the other side, the MA will include initiatives directly addressing adaptation and resilience, such as the use of the shares of proceeds and/or other funds that will be entirely dedicated to policies and actions aimed at facilitating adaptation and resilience activities within the MA.

While adaptation and resilience are included within the initiatives promoted by the MA,[10] other climate related issues, such as loss and damage, are not. The Paris Agreement (PA) includes loss and damage among its priorities throughout Article 8, based upon the Warsaw International Mechanism for Loss and Damage associated with Climate Change Impacts.[11] The MA does not include mechanisms or instruments aimed at minimising and/or addressing loss and damage associated with the adverse effects of climate change. This absence is mainly due to the lack of clarity within the PA relating to similar actions.

---

[10] Even in absence of a corresponding governance model whose design falls beyond the aims of this book.

[11] Paragraph 2 of Article 8 of the Paris Agreement (P2A8PA).

Designing mechanisms for loss and damage appears unfeasible in absence of clearer international legislation to define the scope and rationale of climate related loss and damage. Hence the MA proposal does not explicitly incorporate such aspects within its domain. It is nevertheless possible to foresee the inclusion of loss and damage related initiatives once the international climate legislation will complete the settlement and definition of the corresponding institutional foundations.

The MA timeframe dilemma represents by far its biggest and probably unsolvable limit, at least at the present stage. By defining a NZC target to be reached in a defined period,[12] the MA sets its own timeframe. The system envisaged by the MA can and will only exist as long as there is a mitigation target to be met, hence the achievement of the MA aggregate target will simultaneously represent the achievement of the MA's ultimate goal and the end of its rational for existing. The entire structure, functioning, and governance model of the MA is founded on this ultimate goal and its achievement implies for all the mechanisms, approaches, policies, and actions discussed over this book to lose their reason of being.

At the present stage, where the MA is only an hypothesis and none of its elements have been established and implemented and no certainty can be offered regarding the overall MA's efficacy, this limitation may appear distant in time and negligible in comparison to other, more urgent priorities. Nevertheless, its fundamental impact in the long-term should not be underestimated. Assuming for the MA to be effectively able to provide the numerous, transformational benefits discussed in Chaps. 9 and 10 of this book, it is undeniable that once such benefits will default as a consequence of the target achievement, the corresponding impact on society may be devastating.

Establishing programmes and strategies to counteract, contain, and possibly overcome such potentially devastating evolution of the system should therefore represent a priority for both scholars and policy makers from the very first years of MA implementation. The long-term perspective of the MA and its transformational potential may limit the reliability of studies and evaluations aimed at addressing a 'post-MA society' structure and functioning. Nevertheless, the risks embedded in delayed action addressing the MA timeframe dilemma may become the biggest mistake in the history of climate policy and beyond.

References

Hardin, G. (1968). The tragedy of the commons. *Science 162*(3859), 1243–1248.

Weischer, L., Morgan, J., & Patel, M. (2012). Climate clubs: Can small groups of countries make a big difference in addressing climate change? *Review of European Community and International Environmental Law, 21*(3), 177–192.

[12] See Chap. 5 of this book.

Chapter 12
Conclusions

Abstract Conclusions to this book are organised into three parts: (a) key findings; (b) policy recommendations; (c) conclusions. Set in 22 points, key findings serve as general summary for the Mitigation Alliance (MA) proposal as described in this book. The part referring to policy recommendations attempts to describe the climate policy arena's actors that may most likely adhere to the MA. In doing so, the part briefly summarises the current positions of parties in the climate change negotiations, as well as their general attitude towards climate change. Adopting the 'Earth Statement' as a reference, the final part of this chapter (and of this book) represents a call for immediate action, aimed at avoiding the loss of the priceless value and opportunity offered by the Paris Agreement in ultimately tackling climate change worldwide.

Keywords Key findings • Policy recommendations • Conclusions • Earth statement

12.1 Key Findings

Conclusions to this book are organised into three parts: (a) key findings; (b) policy recommendations; (c) conclusions.

Key findings summarises the narrative of the Mitigation Alliance (MA) proposal in 22 sequential points. Starting from its rationale and origins (Article 6 of the Paris Agreement—A6PA), the findings ultimately result in the exhaustion of the MA's purpose. The 22 identified key findings are summarised as follows:

1. Due to its voluntary nature, A6PA can be adopted to promote experimental climate governance models, such as the one discussed in this book and defined as the MA.
2. MA members can be parties and/or other entities designated/allowed by their reference parties.
3. Following the criteria of A6PA and aligning to the ambitious objectives of the Paris Agreement (PA), MA members define an aggregate target corresponding to a Net Zero Carbon (NZC) emission level by the end of the century.

M. Stua, *From the Paris Agreement to a Low-Carbon Bretton Woods*,
DOI 10.1007/978-3-319-54699-5_12

4. The MA members distribute their NZC target over time according to a set of timetables. To align to A4PA, these timetables will correspond to 5-year periods, while the target's distribution in each timetable will be either equal to or higher than the previous one.
5. The aggregate MA target and distribution over time represent the MA's Plurilaterally Determined Contribution (PDC).
6. The mitigation burden corresponding to the portion of target distributed in each timetable is assigned in quotas among the MA members through the application of a dynamic formula.
7. While quotas represent a mandatory mitigation goal for members assigned with them, some members registering per capita emissions below the MA's aggregate level are exempted from mandatory mitigation in the corresponding timetable.
8. Exempted members will be eventually entitled to increase their emissions to support their development needs.
9. The NZC target and the dynamic quotas system guarantee effectiveness, equity, environmental integrity and transparency of the MA, hence demonstrating its alignment with the PA's principle of Common but Differentiated Responsibilities and Respective Capabilities (CBDR-RC).
10. Members' quotas and their respective climate strategies and policies represent their Members Determined Contributions (MDCs), which correspond to NDCs when members correspond to parties.
11. The MA establishes a single mechanism, defined as the Mitigation Alliance Mechanism (MAM), to eventually certify any mitigation within its domain, including reduction, avoidance and sequestration.
12. The MAM certifies mitigation outcomes throughout the issuance of Certified Mitigation Outcomes (CMO), recognised by the MA members as the only tools to demonstrate their quotas' meeting/achievement.
13. Despite the uniqueness of its certification mechanism, the MA recognises multiple approaches to CMO exchange among MA members and/or stakeholders, including market, non-market and hybrid approaches.
14. The CMO-based system will stimulate a linking process between the different approaches and their convergence towards a single umbrella defined as an Inclusive Offsetting Scheme (IOS).
15. A multi-level, hybrid model of governance, where overarching authorities similar to those established by Bretton Woods are counterbalanced by high degrees of freedom for MA members and stakeholders, is set up.
16. The MA recognises carbon as an exhaustible natural resource, the consumption of which produces carbon emissions and the conservation of which constitutes mitigation.
17. The demand for carbon conservation, represented by the MA target and corresponding quotas, provides carbon with an actual value, making it a commodity.
18. Carbon conservation in form of CMOs represents the representative currency for the commodity.

19. Combined with the MA governance model, the identification of a commodity and corresponding representative currency transforms the MA into a Low Carbon Bretton Woods system.
20. The implementation of the MA ensures multiple benefits for its members and stakeholders that, based upon the PA recognition of the value of voluntary mitigation outcomes, can be organised into three categories: (a) environmental benefits; (b) economic benefits; and (c) social benefits.
21. The combination of the MA's multiple benefits achievable through full MA implementation can lead to radical industrial innovation, societal change and environmental transformation, and decoupling human growth and carbon emissions while leading to a deep decarbonisation of the world.
22. Whatever action will be taken within the MA, its system will collapse with the achievement of its main purpose, the NZC.

12.2 Policy Recommendations

As discussed in Chap. 11 of this book, the efficacy of the Mitigation Alliance (MA) will heavily rely upon its ability to attract and establish a *critical mass of relevant members*. The overall architecture of the MA, as well as its close relationship with the Paris Agreement (PA) and the United Nations Framework Convention on Climate Change (UNFCCC), encourages signatories of the PA to become MA members. It is therefore possible to affirm that, in most cases, it is necessary to view parties as ideal 'cooperative entities' within the MA framework. Nevertheless, the opportunity for MA membership to include entities other than parties provides the basis for a multi-layered and more inclusive governance organisation compared to those traditionally characterising the international climate regime. It is therefore possible to foresee a new and more active policy role for public and private stakeholders other than UNFCCC parties.

Local authorities, such as provinces, states belonging to federations, and cities (also in the form of coalitions) are most likely to become primary actors of the MA. Evidence from climate governance demonstrates how effective climate awareness at local level may result in much more advanced and structured action compared to parties (national) level (Hamin and Gurran 2009; Puppim de Oliveira 2009; Gallaraga et al. 2011; Bae and Feiock 2013). Significant achievements in climate governance have arisen at sub-national level, including some of the most advanced carbon markets. For instance, cases like California (Krebs 2011; Bedsworth and Hanak 2013) or some Canadian provinces represent well-established climate governance innovations, while experiments in big urban areas of Asia have provided interesting insights into the potential development of effective local policies which address the different needs and priorities resulting from the climate change challenge (Takao 2014; Chen 2015; Yoon et al. 2016).

Moreover, international coalitions of cities and local authorities represent key actors in the move towards an ambitious agreement like the PA, consolidating their role as primary actors in the field of international climate policy.[1] More generally, we have witnessed a significant rise of civil society engagement in the climate debate in recent years, with an increasing number of local communities, movements, firms, and industries, as well as varied coalitions of interest, converging towards a shared effort to effectively implement a reliable climate strategy worldwide (Bernauer et al. 2016; Hale 2016; Jacobs 2016). Therefore, successful establishment of the MA calls for a direct and full involvement of such stakeholders in all its processes from the very beginning. The opportunity for entities other than parties to become members of the MA and the institution of measures to stimulate participation of other key stakeholders must be a priority in the implementation of the MA.

The MA proposal is likely to appeal to a multitude of parties, with various rationales for supporting the initiative. Some in particular are more likely to recognise the significance of the opportunities offered by the MA and are thus likely to be first movers in its implementation process. It is worth considering these in more detail, along with the most recent positions taken by the different parties at the climate negotiation table, as well as individual climate policies and stances for future climate-related strategies. In what follows a simplified socio-geographical clustering is adopted for clarification.

For instance, coalitions of parties that took the lead in the process towards an ambitious, international agreement on climate change may more easily recognise the significance of the MA and therefore prove to be 'first movers' in its implementation. Entities such as the Alliance of Small Island States (AOSIS) and the Independent Association of Latin America and the Caribbean (AILAC) pushed for the strongest commitment for an uncompromised agreement in Paris (Chang and Soruco Carballp 2011; De Agueda Corneloup and Mol 2014; Hoad 2016) despite contributing only a small percentage of global emissions. Such entities maintain advanced positions in the overall climate debate, often managing to 'pave the way' for other entities to follow. Interest from such coalitions may represent an important starting point for the MA.

Other groups, whose actions and strategies were often aligned with their own 'private' interests but nonetheless managed to overcome their differences to converge towards the Paris Agreement (Rajamani 2016), such as the Umbrella Group, the Group of 77 plus China (G77+China), or the Organisation for Economic Co-operation and Development (OECD), may find in the equity of the MA proposal and, more importantly, in its variety of private and excludable benefits, significant reasons to join. The European Union (EU), which represents the most formal entity among these groups, requires some additional considerations. On the one hand, the EU has a long tradition as primary actor in climate negotiations and represents a pioneering entity for the deployment of carbon markets on a large scale. On the other hand, internal tensions and contradictions that have been threatening the EU in recent years, as well as the need to simultaneously take into consideration the political stances and economic priorities of 28 different member states, have now

[1] See also Sect. 2.3 of Chap. 2 of this book.

weakened its determination and relevance as policy actor in the climate change context (Jordan et al. 2012; Skovgaard 2014; Benjamin et al. 2015).

In the case of the above listed groups it is more likely for their members to adopt an 'individual' approach towards the MA, with single parties that already showed 'advanced' climate-related positions more likely to embrace the idea. Such parties may be identified in each of the above-mentioned groups and a concise and incomplete list of them would consist of entities from all over the world and from varied socio-economic realities. For instance, the list could include parties that showed commitment and ambition at the negotiating tables, as well as countries that promoted national innovations in climate governance.

An incomplete list of parties that showed strong political commitment and/or vision for innovation would include, for instance, France and Brazil, who closely worked for the successful implementation and inclusion of Article 6 in the PA (A6PA) (Marcu 2016); Canada, who strongly promoted the adoption of the 1.5 °C target within the PA (Prystupa 2015); Japan, who led the very first initiatives for the design of PA implementation tools (Koakutsu 2016); China, who encouraged a global agreement through a proactive attitude in conjunction with the United States of America and implemented innovative climate governance strategies (Huang and Bailis 2015; Jiang et al. 2016; Swartz 2016; Wang and Zhang 2016); and Germany, first mover within the EU for climate and clean energy initiatives (Pregger et al. 2013; Buchsbaum 2016).

The Least Developed Countries (LDC) should also recognise the importance of adhering to an MA proposal which is fully supportive of their situations vis-a-vis Common But Differentiated Responsibilities and Respective Capabilities (CBDR-RC). By joining the MA they may eventually become the primary recipients of MA-related benefits without incurring unfair costs. Yet these parties and countries may have some prejudice towards a proposal that, at first sight, seems to have some similarities with unpopular traditional, liberal policies.

Finally, some entities that relied heavily on fossil fuels for their income, such as the Organization of Petroleum Exporting Countries (OPEC), will likely consider themselves potential losers within the MA, and are unlikely to adhere to it during its early stages. For other parties it is more difficult to anticipate their response and interest in the MA. The United States of America (USA), for instance, who engaged proactively in climate negotiations in recent years, may radically change direction as a result of the presidential election at the end of 2016.

Russia, moreover, which often maintained a 'neutral' position in the climate change debate and, more importantly, has always prized full sovereignty as one of its key priorities, may simply have no interest in getting directly involved in an ambitious and challenging proposal such as the MA.

Finally, a country like India, which may extensively benefit from adhering to the MA, has often maintained an ambiguous approach to climate change (Jaeger and Michaelowa 2016), and may therefore refuse to 'take the lead' with a challenging MA framework (Thaker and Leiserowitz 2014).

In conclusion, the MA has potential to effectively attract a critical mass of relevant members. It also provides the opportunity to further the unique geopolitical shifting process that several scholars have identified as a key driver of climate negotiations in recent years. In other words, the MA may effectively become the 'policy context' where the twentieth century heritage is replaced by a twenty-first century multipolar international regime (Dunford and Yeung 2011; Timmons Roberts 2011; Papa and Gleason 2012; Hochstetler and Milkoreit 2014).

12.3 Conclusions

To *solve* has represented the fundamental objective of human intellect since its very early application. Human beings developed intellect with the primary aim of *solving* the multiple issues that challenged and threatened them over time. The ability to *solve* has maintained the same, fundamental role in human existence over the centuries, becoming the primary driver for the evolution of human intellect. Philosophers, magicians, scientists, and scholars have always been driven by the human need and wish to *solve*.

Climate change is arguably the biggest current threat to humanity and has attracted thousands of minds dedicated to *solving* it. The 'Earth Statement',[2] written by 17 of the world leading scientists[3] and supported by more than 100 internationally renowned personalities, constitutes the ultimate attempt to *solve* the climate issue. Identifying eight essential elements of climate action to safeguard human development, the 'Earth Statement' represents a complex yet feasible introduction to what is required to *solve* the climate change issue.

With all its limits, this book and the proposal it contains have been written with the 'Earth Statement' in mind, attempting to outline one pathway for its successful application in the global context of today. In other words, the hope is that this book can help *solve* the climate dilemma and its accompanying social justice issues worldwide.

Undoubtedly overambitious in its aims, this book has been written with full consciousness of the need for ambition and vision to address the tragedy of commons and to move towards the *common good* as the only resource capable of providing for the safety of ourselves and the world that surrounds us.

---

[2] Earth Statement: http://earthstatement.org/statement/.

[3] Johan Rockström (Stockholm Resilience Centre), Guy P. Brasseur (Max Planck Institut für Meteorologie), Ottmar Edenhofer (Mercator Research Institute), Sir Brian Hoskins (Grantham Institute), Pavel Kabat (International Institute for Applied Systems Analysis), Mario J. Molina (Centro Mario Molina), Jennifer Morgan (World Resources Institute), Nebojsa Nakicenovic (International Institute for Applied Systems Analysis), Carlos Nobre (Instituto Nacional de Pesquisas Espaciais), Veerabhadran Ramanathan (Scripps Institution), Jeffrey Sachs (Earth Institute), Hans Joachim Schellnhuber (Potsdam Institute), Peter Schlosser (Earth Institute), Youba Sokona (South Center), Leena Srivastava (TERI University), Lord Nicholas Stern (Grantham Institute), Guanhua Xu (Chinese Academy of Sciences).

References

Bae, J., & Feiock, R. (2013). Forms of government and climate change policies in US cities. *Urban Studies, 50*(4), 776–788.

Bedsworth, L. W., & Hanak, E. (2013). Climate policy at the local level: Insights from California. *Global Environmental Change, 23*, 664–677.

Benjamin, C., Cadoret, I., & Hubert, M.-H. (2015). The European climate policy is Ambitious: Myth or reality? *REP, 125*(5), 731–753.

Bernauer, T., Gampfer, R., Meng, T., & Su, Y.-S. (2016). Could more civil society involvement increase public support for climate policy-making? Evidence from a survey experiment in China. *Global Environmental Change, 40*, 1–12.

Buchsbaum, L. (2016). Germany's Energiewende at a new turning point. *Power, 160*(3), 1–8.

Chang, C. C., & Soruco Carballp, C. F. (2011). Energy conservation and sustainable economic growth: The case of Latin America and the Caribbean. *Energy Policy, 39*, 4215–4221.

Chen, V. Y. (2015). The role of urban green infrastructure in offsetting carbon emissions in 35 major Chinese cities: A nationwide estimate. *Cities, 44*, 112–120.

De Agueda Corneloup, I., & Mol, A. P. J. (2014). Small island developing states and international climate change negotiations: The power of moral "leadership". *International Environmental Agreements, 14*, 281–297.

Dunford, M., & Yeung, G. (2011). Towards global convergence: Emerging economies, the rise of China and western sunset? *European Urban and Regional Studies, 18*(1), 22–46.

Gallaraga, I., Gonzalez-Eguino, R., & Markandya, A. (2011). The role of regional governments in climate change policy. *Environmental Policy and Governance, 21*, 164–182.

Hale, T. (2016). "All hands on deck": The Paris Agreement and nonstate climate action. *Global Environmental Politics, 16*(3), 12–22.

Hamin, E. M., & Gurran, N. (2009). Urban form and climate change: Balancing adaptation and mitigation in the U.S. and Australia. *Habitat International, 33*, 238–245.

Hoad, D. (2016). The 2015 Paris Climate Agreement: Outcomes and their impacts on small island states. *Island Studies Journal, 11*(1), 315–320.

Hochstetler, K., & Milkoreit, M. (2014). Emerging powers in the climate negotiations: Shifting identity conceptions. *Political Research Quarterly, 67*(1), 224–235.

Huang, Y., & Bailis, R. (2015). Foreign policy 'trilemmas': Understanding China's stance on international cap-and-trade. *Climate Policy, 15*(4), 494–516.

Jacobs, M. (2016). High pressure for low emissions – How civil society created the Paris climate agreement. *Juncture, 22*(4), 314–323.

Jaeger, M. D., & Michaelowa, K. (2016). Global climate policy and local energy politics: Is India hiding behind the poor? *Climate Policy, 16*(7), 940–951.

Jiang, J., Xie, D., Ye, B., Shen, B., & Chen, Z. (2016). Research on China's cap-and-trade carbon emission trading scheme: Overview and outlook. *Applied Energy, 178*, 902–917.

Jordan, A., van Asselt, H., Berkhout, F., Huitema, D., & Rayner, T. (2012). Understanding the paradoxes of multi-level governing: Climate change policy in the European Union. *Global Environmental Politics, 12*(2), 43–66.

Koakutsu, K. (2016). The Joint Crediting Mechanism (JCM) as an example of Article 6.2 activities. *Carbon Mechanisms Review, 1*, 12–15.

Krebs, M. (2011). Lessons of climate change, stories of solutions. California: Government leadership. *Bulletin of the Atomic Scientists, 67*(1), 60–63.

Marcu, A. (2016, January). *Carbon market provisions in the Paris Agreement (Article 6)* (CEPS Special Report, 128).

Papa, M., & Gleason, N. W. (2012). Major emerging powers in sustainable development diplomacy: Assessing their leadership potential. *Global Environmental Change, 22*, 915–924.

Pregger, T., Nitsch, J., & Naegler, T. (2013). Long-term scenarios and strategies for the deployment of renewable energies in Germany. *Energy Policy, 59*, 350–360.

Prystupa, M. (2015, December 7). Canada shocks COP21 with big new climate goal. *National Observer*.

Puppim de Oliveira, J. A. (2009). The implementation of climate change related policies at the subnational level: An analysis of three countries. *Habitat International, 33*, 253–259.

Rajamani, L. (2016). Ambition and differentiation in the 2015 Paris Agreement: Interpretative possibilities and underlying politics. *International and Comparative Law Quarterly, 65*, 493–514.

Skovgaard, J. (2014). EU climate policy after the crisis. *Environmental Politics, 23*(1), 1–17.

Swartz, J. (2016, March). *China's national emissions trading system – Implications for carbon markets and trade* (Issue Paper 6). International Centre for Trade and Sustainable development – ICTSD.

Takao, Y. (2014). Policy learning and diffusion of Tokyo's metropolitan cap-and-trade: Making a mandatory reduction of total CO2 emissions work at local scales. *Policy Studies, 35*(4), 319–338.

Thaker, J., & Leiserowitz, A. (2014). Shifting discourses of climate change in India. *Climatic Change, 123*, 107–119.

Timmons Roberts, J. (2011). Multipolarity and the new world (dis)order: US hegemonic decline and the fragmentation of the global climate regime. *Global Environmental Change, 21*, 776–784.

Wang, X., & Zhang, S. (2016). Exploring linkages among China's 2030 climate targets. *Climate Policy*. doi:10.1080/14693062.2015.1124752.

Yoon, I., Lee, Y., & Yoon, S. K. (2016). An empirical analysis of energy efficiency measures applicable to cities, regions, and local governments, based on the case of South Korea's local energy saving program. *Mitigation and Adaptation Strategies for Global Change*. doi:10.1007/s11027-016-9702-3.

A Low Carbon Bretton Woods: A Legal Critique

Peter Zaman

Context

There has been much analysis of the Paris Agreement and what it means in terms of the international response and framework for addressing climate change. From a legal perspective, much of the analysis has focused on whether it provides sufficient binding targets to successfully achieve the desired ambition of ensuring the impact of adverse global warming is confined to a less than 2 °C change in average global temperature (the "**Below 2 °C Target**"). This has focused attention on the Nationally Determined Contributions ("**NDCs**") of the respective Parties to the Paris Agreement and whether the Paris Agreement provides the institutional framework to enforce those national outcomes. The short answer to this is that the Paris Agreement does not have the direct enforcement mechanism that enables the United Nations to take a country to task and impose penalties for non-compliance with its NDCs. Instead it relies on each of the Parties voluntarily co-operating in the collective goal of achieving the Below 2 °C Target. Some argue that this is a weakness in the Paris Agreement whilst others recognise that it was a necessary outcome to deliver a universally acceptable truly international treaty on climate change with contributions from all Parties.

Described as a 'bottom-up' approach (as compared to the 'top-down' approach of the Kyoto Protocol) the Paris Agreement, in truth, includes a mixture of binding and non-binding obligations. The author of this book has already summarised those various obligations in Chap. 2 of this book. What is however clearly missing as a binding obligation on the part of each of the Parties to the Paris Agreement, is an obligation to achieve specific greenhouse gas emission reduction targets (hereafter referred to as 'carbon' mitigtion targets) over a specific period. The Paris Agreement instead allows for the Parties to meet the Below 2 °C Target by ratcheting up their ambition to reflect a peaking of global emissions, expressed via their NDCs, along a timeframe guided by the "best available science, so as to achieve a balance between anthropogenic emissions by sources and removals by sinks of greenhouse gases in

M. Stua, *From the Paris Agreement to a Low-Carbon Bretton Woods*,
DOI 10.1007/978-3-319-54699-5

the second half of this century".[1] Equally, the Paris Agreement leaves the Parties to decide for themselves as to the method by which they intended to achieve their respective contributions towards the global effort of meeting the Below 2 °C Target (e.g. whether that is by adopting a cap and trade mechanism, a carbon tax or through other non-market based policy measures or whether that is a singular target or one shared with other like-minded countries).

However, the underlying but unstated position in the Paris Agreement, reflected via the guidance that science presently affords, is that, the Below 2 °C Target is only achieved if we reduce our collective greenhouse gas ("**GHG**") emissions to less than a specific number of tonnes of CO_2 equivalent within a specified timeframe of 2100, with a peaking of global greenhouse gas emissions much earlier than that. As stated by the book's author in Chap. 9, this equates to a global carbon budget.

The "Intergovernmental Panel on Climate Change ("**IPCC**") has already indicated that, to have a greater than 66% chance of keeping within the 2 °C goal by 2100, only another 1000 $GtCO_2$ of CO_2 (our global "**Carbon Budget**") can be emitted between 2011 and 2100 (even less to be within the Below 2° Target)."[2] Therefore, what the NDCs must do, over time, is through their collective aggregate ambitions identify a pathway that keeps us within that stated budget of less than 1000 $GtCO_2$ of CO_2. The challenge for a bottom-up treaty such as the Paris Agreement, is to create a facilitative or enabling environment in which such a carve-up of that Carbon Budget is possible. Currently, any such allocation of a Carbon Budget in the Paris Agreement is merely implicit and buttressed with qualifications expressed though principles such as "common but differentiated responsibilities and respective capabilities, in the light of different national circumstances" ("**CBDR&RC**"). Alternatively, as viewed by the author of this book as "burden sharing", the idea of the Carbon Budget, in the context of the Paris Agreement, is particularly challenging given the strong force which led to a shift away from the Kyoto Protocol's efforts of allocating carbon caps on Annex 1 countries to the Kyoto Protocol.

Introduction

It is in the context of this background that the author of this book introduces his interpretation of Article 6 to the Paris Agreement ("**A6PA**") as the facilitative or enabling environment in which to effectively manage that Carbon Budget. In doing so, we are introduced to the concepts of the Mitigation Alliance ("**MA**"), Mitigation Alliance Mechanism ("**MAM**"), Plurilaterally Determined Contribution ("**PDC**"), Net Zero Carbon ("**NZC**"), Certified Mitigation Outcome ("**CMO**") etc. To implement these concepts, we are further introduced to the idea of a governance structure that involves international bodies in the form of the Mitigation Alliance General

[1] Article 4(1), Paris Agreement.

[2] Zaman, P.; Rock, N.; Hedley, A.; Smokelin, J. (2016) The Paris Agreement: Leading the Pathway to a Low Carbon Economy. 16 *Platt's Energy Law Report* 4, *LexisNexis A.S. Pratt*, p. 131–149.

Assembly ("**MAGA**"), Mitigation Alliance Target Fund ("**MATF**"), Mitigation Alliance Target Fund Executive Board ("**MATF-EB**"), Certified Mitigation Outcome Bank ("**CMO Bank**") and the Certified Mitigation Outcome Exchange Board ("**CMO-EB**").

The MA Proposal

At the heart of the author's proposal is the idea that, a number of Parties to the Paris Agreement, with an intention to reflect a higher ambition than merely that of their individual NDCs, may voluntarily from a club of Parties as permitted within the scope of A6PA. Such club or MA, as formed, should be in a manner that is, consistent with the Paris Agreement overall. The individual ambition of the club's members is to be quantified in terms of GHG (i.e. carbon) emissions' mitigation (or simply mitigation) and an agreed timetable, established as an aggregated or combined carbon budget and, finally carved up as an individual member's budget (in the forms of quotas).

These quotas are not in the form of carbon allowances (as traditionally seen in cap and trade models or in Annex B of the Kyoto Protocol) but as assigned shares of the mitigation burden or mitigation objectives of a collective goal (see Chaps. 5 and 7). As stated in Chap. 5, "having calculated the per capital level of GHG emissions, the resulting values will be used to establish which members will be responsible for the share of mitigation target corresponding to the given timetable". The calculations are driven by the idea of NZC, as adjusted by the Parties equity considerations such as CBDR&RC and by thresholds reflecting a lower emission per capita of the respective member. Quotas are therefore, 'weighted' by the MA member's respective share and are time sensitive. If the quotas reflect a share of the overall target, the PDC represents the redistribution of that target in a manner that collectively ensures consistency with those members aggregate Paris Agreement commitments as represented under their NDCs.

The MA proposal allows evidence, of the success of a member in achieving its respective quota, to be measured by a common unit of calculation, called the CMO. However, the CMO is not a carbon offset, an allowance unit within a cap and trade system nor a form of tax credit. The MAM, a certification system modelled on the CDM but broader in that it is wide enough to capture any form of mitigation activity (e.g. ETS, offsetting, REDD+, afforestation, reforestation), is the source of the unit representing the successful achievement of the mitigation objective, the CMO. The CMO unit, represents "the totality of mitigation outcomes required for meeting the MA member's quotas and aggregate NZC target…".[3] The author envisages the use of the CMO, as the tool to enable indirect linking to occur between members within the MA.

[3] See Sect. 7.5.

The CMO is also described as the representative currency of the MA because it is the unit of measurement for documenting the achievement of quotas. Under its terms, the MA will require its members to hold a specific quantity of CMOs as reserve assets (by analogy to the special drawing rights ("**SDR**") of the IMF). It will be the MA member's prerogative to establish models to convert their cap and trade allowances or offsets etc. into CMOs based on exchange rates. It is further envisaged that MA members who have met their target can bank a corresponding quantity of CMOs.

The link therefore, between the MA mitigation target and CMO is, for the book's author, an essential prerequisite to allow carbon to be treated as an asset that is capable of being recognised at public law within the MA as well as, potentially, at private law. For this, the author compares the remaining carbon emissions capable of being emitted within the Carbon Budget as an exhaustible, quantifiable, direct, excludable benefit for the MA members akin to a commodity that is scarce and valuable. If carbon, is an exhaustible natural resource then "mitigation equates to the conservation of the resource".[4] As a consequence, CMOs represent the unit of measurement "for quantifying the MA's carbon conservation objectives, as expressed by its mitigation target". Therefore, within a MA member's target, the greater its carbon conservation, after meeting its mitigation target, the more it will have available to it by way of CMOs which it can then bank or add to its reserves for future use. Based on the strength of its reserves, the tradeable basis of the CMO will enable countries to sell or if needed, further acquire CMOs. The CMO therefore becomes a reserve asset under the MA.

As expressed by Professor Levi, any "reserve asset which is internationally traded will offer an efficiency gain by equating economic costs of alternative policies for correcting international payment imbalances, provided all currencies are convertible into that reserve asset."[5] The efficiency of a tradable quota system is that it drives the marginal cost of reducing emissions to equality. The Bretton Woods Agreement "achieved the efficiency property …via the SDR …".[6] This is where the analogy to the SDR, to the CMO, begins to arise.

Just as an IMF member's SDR enables it to settle trade account deficits, an MA member that has not satisfied its quota should be able to acquire CMOs from those members who have surplus CMOs. Like the original Bretton Woods system, if all MA members had to value their carbon equivalent in the form of a CMO, the CMO establishes itself as that member's reference currency. This in turn allows "MA members to compare each other's economies by comparing their currencies' exchange rates against CMO (Mehrling 2013)."[7] In this instance, under the MA approach, an MA country's mitigation effort is measurable against its mitigation

---

[4] See Sect. 9.2.

[5] Levi, M. D. (1991). Bretton Woods: Blueprint for a greenhouse gas agreement. *Ecological Economics*, 4, 253–267, at p. 259.

[6] Levi (1991), at p. 259.

[7] Section 9.4.

objective, in the form of quotas, and its account of success or failure will be reflected in its CMO reserve account position held with the MATF.

However, a distinction is drawn between the CMO as the MA's 'representative currency' (i.e. for settlement of obligations recognised and arising under the MA between members) and carbon as a 'reference commodity'. Today, SDRs are a right to draw on the component convertible currencies of the basket of SDR currencies held with the IMF rather than a currency in its own right. As originally created, the value of the SDR was fixed at 0.888671 g of fine gold which was also worth 1 USD.[8] By analogy, the proposal is for the a CMO to be backed by 1 tonne CO_2 equivalent and have its value determined by the price a country would have pay within the MA for a tonne of CO_2 equivalent. As a country's carbon conservation/mitigation efforts increase, it will have more CMOs to bank or sell. If a member country is failing in its mitigation efforts, under the terms of its membership, the MATF could require that country to purchase CMOs from other countries that have reserves, thereby ensuring a demand for CMOs amongst members and, depending on the constitutional structure of the MA, a demand from private sector actors.[9] This book's author contemplates that under the MA system, the private sector would be engaged in the mitigation process either at national or international level and would transact in CMO's, eventually extended to non-parties when their involvement was permitted by their reference Parties under the MA.

The Governance Framework to Support the MA Proposal

Inspired by the governance framework for the original Bretton Woods Agreement, the author proposes several institutions to support the activities, policies and implementation of the MA proposal.

These institutions are contemplated to have, at all times, recognition (in some cases designation) by the United Nations Framework Convention on Climate Change ("**UNFCCC**") and/or the Paris Agreement. Therefore, although the MA is essentially a carbon club that acts independent of the UNFCCC and the Paris Agreement, for it to be truly effective, there are components of the club's tools that require the endorsement of the relevant bodies to the Paris Agreement or the UNFCCC.

For example, in the role imagined for the CMO-EB and its interaction with the UNFCCC and the Conference of the Parties to the Paris Agreement ("**CPA**") regarding the MA's PDC compliance, it will require recognition as the MA's agent to the UNFCCC as well as a status to enable it to access other international organisations.

---

[8] Today, the value is pegged against the value of a basket of SDR currencies.

[9] Notably, SDRs are not tradable among private sector participants under the IMF's Articles of Agreement.

The author of the book envisages the roles of the various bodies to be as follows (see further Chap. 8):

Institution	Role
MAGA	The founding authority and ultimate regulator of the MA. It sets up the MA and the three other institutions and sets the legislative norms by which the MA, its institutions and members will operate. The MAGA will have equal representation of all MA members and determines the criteria for membership and guaranteeing the consistency of the MA with A6PA and the Paris Agreement
MATF (analogous to the IMF)	Its role is to define and distribute the quotas under the MA as well as to manage the compliance of the accounting rules by its members. It establishes and manages the demand for CMO and is managed through a Board of Governors and the MATF-EB. It sets the CMO reserve requirements for its members and manages the accounting risks of the MA system. The MATF-EB manages and operates the MA registry which is used as a tool (e.g. to avoid double counting of CMOs) within the MA system. Specifically, the MATF-EB is responsible for enforcing the regulatory obligations of MA members. As part of the MA risk management objective, the MATF may be empowered to apply a member country 'pseudo-rating' (e.g. based on its performance or likelihood in achieving its quotas or targets or upon the strength of its CMO reserve balances) or to apply a penalty (e.g. by requiring non-performing members to acquire CMOs from other surplus member countries)
CMO Bank (analogous to the World Bank)	Representation at the bank will be in proportion to the relevant member country's share of the quotas. Its mandate will include, increasing sustainable development among MA members, support the meeting on the MA's ambitions (e.g. by being responsible for CMO certification and issuance), operating the MAM, if appropriate—being the point of liaison with the private sector (by analogy to the IFC)
CMO-EB (inspired by the WTO)	Their role is to oversee the implementation of the convertibility of a member's domestic mitigation activity into CMOs. It will also be able to resolve any disputes between MA members arising from who are trading with each other or who's national policies conflict with the MA's principles and norms, on which another affected or interested member has raised a complaint. The exchange rate that enables one member's mitigation effort to be converted to a CMO, as distinct from the mitigation effort of another member, would be established and set by the CMO-EB

Challenges and Limitations Many superlatives come to mind when first reading the MA proposal. Among them, the foremost, is 'ambitious'. The use of A6PA to bridge the gap between what the Paris Agreement did do and what it should have further done, requires both imagination and rationalisation. The author does not shy away from the task of addressing most of the challenges that follow from the MA proposal and provides much to think about in terms of supporting argument, evidence and theory. However, what follows in this chapter are reflections and comments from the perspective of a legal practitioner who, from experience, is only too conscious of the further challenges that nonetheless remain with the MA proposal. The discussion below does not address these challenges in any particular order of significance or difficulty and nor is it an exhaustive list. However, these challenges are the ones that leapt out to this practitioner upon his consideration of this book's author proposal.

Conceptual Leap

The author's interpretation of A6PA rests on his view of it as a single cohesive article within the Paris Agreement where its constituent elements, provided by paragraphs (1) to (8), are intertwined with each other. This enables this book's author to extract ten overall principles (see Chap. 4) supported by A6PA that are the 'founding elements' that allow the author to link Article 6 to cooperative game theory and climate club concepts. Elements of each of these ultimately help form the MA proposal.

For those who were closely involved with the Article 6 negotiations during COP21, A6PA was described as nothing less than a 'miracle'. However, it's appearance in the drafting text during the near final hours of the Paris meeting suggest that this was very much a last-minute compromise proposal rather than one which had the opportunity for detailed consideration and discussion among the Party negotiators. Although the origin for many of the provisions within Article 6 can ultimately be traced back to various discussion texts (e.g. links between the Brazilian proposal of 10 November 2014 in the lead up to COP 20 in Peru, the EU-Brazil submission of 8 December 2015, etc.),[10] it is far easier to view A6PA as one where rather vague, broad concepts were thrown at the Party negotiators so that each negotiating group could find in A6PA something that they were each seeking. For example, for those Parties who did not support market mechanisms, they took comfort in Article 6(8). Similarly, for those who wished to utilise the depth of knowledge established through the CDM, they took comfort in Articles 6(4)–(7). Last, but not least, for those who were seeking market mechanisms they took joy in the vagueness of ITMOs under Articles 6(1)–(3).

It is therefore easier to contemplate A6PA as a cobbled-together last-minute compromise that allowed each Party negotiator to take comfort that their respective position has been reflected in A6PA. In the version of the Paris Agreement negotiating text of 10 December 2015, some of the provisions that eventually made their way into Article 6(4)–(7) appeared as a stand-alone Article 3 ter, whereas provisions resembling the eventual text of Article 6(1)–(3) appeared as a subpart of Article 3(20)–(21). In short, when interpreting the overall meaning of A6PA, the views of the Parties that argued for non-market mechanism under paragraph 8 are unlikely to match those who argued for ITMOs under paragraphs (1)–(3). This means that taking an interpretative position of A6PA, that views paragraphs (1)–(8) as a joined-up, inter-linked, harmonious provision and which then allows for the ten principles to be extracted as a foundation for a single overall concept of expressed as the MA proposal, does require a conceptual leap of faith in terms of a reader's perspective on A6PA.

The need to take this interpretative leap of faith to enable the foundations for the MA to exist therefore, reflects its first challenge. As we have learnt from the history

[10] Marcu, A. (2016) Carbon Market Provisions in the Paris Agreement (Article 6). *CEPS Special Report*, No. 128/January 2016.

of the Kyoto Protocol, last minute provisions, as was the case of Article 12 (CDM) of the KP, can often be very successful where there is a will to drive it forward. This too may be the case for A6PA. With the passage of time, change of the guard among Party negotiators, the presence of political will to establish something akin to the MA proposal and a sufficiently large buy-in (see the discussion immediately below), the interpretative leap taken by this book's author may become simply a mainstream view of A6PA. That said, with COP21 still fresh in the minds of many, this interpretation, may today seem ambitious.

Nonetheless, if there was political willingness to support the MA proposal among a 'coalition of the willing', then the author of this book has provided an insightful and inspirational route that would enable them to set out on a journey that is not merely speculative and wishful.

Critical Mass

To implement the MA proposal, it is obviously right to draw parallels to the institutions and framework that was borne out of the Bretton Woods Agreement. However, to continue with the analogy with Bretton Woods and its institutions, also requires a recognition of some of the characteristics and features of history that enabled the creation of a multilateral financial system. As recognised by Broughton (2004),[11] "the International Monetary Fund was forged from failure". Boughton highlights the impact of the failure of the Paris peace conference in 1918 (at the end of the first World War), the great depression between 1929–39 and the rebuilding effort of the Second World War among many of the influencing factors that shaped the views of the founding countries of the Bretton Wood Agreement. On any argument, such events are momentous and extreme.

However, it was necessary for such extreme events to exist to build the coalition of the countries willing and necessary to force though agreement on the Bretton Woods system. These countries recognised that the sacrifice of part of their economic sovereignty was more important, in the interest of breaking down barriers to global trade, than allowing the nationalistic, protectionist measures that flourished during and because of the great events highlighted above. An ideal of the greater public good existed. That willingness eventually led to 44 countries meeting in 1944 to thrash out the Bretton Woods Agreement. However, that window of willingness was arguably, short-lived. "The importance of Bretton Woods as a wartime event was that it took advantage of a window of opportunity to create a multilateral financial system. Both before and after the war, there was too much suspicion and national self-interest for such a sweeping agreement to be possible".[12]

[11] Boughton, J.M. (2004) The IMF and the Force of History: Ten Events and Ten Ideas That Have Shaped the Institution. *IMF Working Paper*, WP/04/75, at p. 2.

[12] Boughton (2004), at p. 7.

A similar challenge would apply to the MA proposal. History may judge that the timing of the Paris Agreement, where China, the US, Europe and the critical mass of the other countries were able to overcome past differences to forge a common path forward, could have been the pinnacle for climate change cooperation in this decade. Further, deeper and more (sovereignty sacrificing) cooperation among several counties seeking to implement the MA proposal and willing to make the necessary sacrifices and take on the founding and funding costs of the MA institutions is not impossible. However, without a certain critical mass of founding members (who also should have significant mitigation obligations under the Paris Agreement), the analogy to the Bretton Woods system fails. In theory, carbon clubs can be formed without a significant critical mass, but the limitation of the MA proposal is that its success hinges on the club's ability to grow and attract new members that are willing to sign up to the MA proposals rules, norms and governance framework as was the case under Bretton Woods.

However, as the Bretton Woods process shows, not all Parties to the Paris Agreement must be involved in its initial set up. The membership of the IMF, for example, has progressively expanded over the years.[13] This is because of the open-doors inclusive policy adopted in the Bretton Woods framework and the striking of an appropriate balance between incentives for joining compared to the price a non-member paid for being outside of the system. Arguably, a country's decision to join the multilateral institutions forged by the Bretton Wood Agreement are driven by its own specific drivers (such as the need for financial support by many of former Soviet Union countries after the fall of the Berlin wall in 1989) and not all countries will be driven by the same desire to join. However, the common theme that exists in all country decisions to join are driven by economic considerations.

The importance of the economic drivers to a country or to the politics that influences and prioritises that country's policies are hard to compare with the significance of a country's priorities for climate change. It is probably, fair to say, that short of an environmental catastrophe, a country's political agenda is unlikely to reflect neither the priority nor the urgency of the climate change challenge. The historic U.S. elections of 2016 were symbolic in how little importance was placed by a significant proportion of its electorate on issues relating to climate change. Therefore, one of the biggest obstacles of the MA proposal is whether, the willingness and leadership required to establish the MA proposal will ever flourish within enough Parties to the Paris Agreement who's GHG mitigation and climate change obligations are important and significant enough to establish a large enough club that will attract future members. Even where they do, the question is can it attract enough members *within time* to nonetheless ensure successful achievement of the Below 2° Target?

[13] In 1945 it was founded with only 29 country members, by 1969 had 115 members, by 1994 had 172 members and today it has 189 members (see https://www.imf.org/external/np/sec/memdir/memdate.htm)

The World Bank's MV Concept

The MA proposal is not framed in the language of linking of carbon markets. Although it draws heavily on carbon club literature and recognises the use of the MA as facilitating a convergence towards a single market (see Sect. 7.4), it does not spend a lot of time discussing the various linking models proposed in various carbon club literature. This literature has recently been highlighted again within the World Bank Group's own post-Paris Agreement work as part of its carbon pricing initiative[14] and as reflected in its Networked Carbon Markets ("**NCM**") concept.

The NCM seeks to "enable comparison of different mitigation actions and trade across different mitigation outcomes in a way that is: inclusive; transparent; efficient; and has environmental integrity".[15]

There are, therefore, parallels between the issues the NCM is seeking to address and the MA proposal. In particular, "a key element of the NCM initiative is a risk-based approach to assessing climate change mitigation value of heterogeneous mitigation actions".[16] Under the NCM process, mitigation value ("**MV**") is described as "a measure of the effectiveness of mitigation efforts, or put another way, the mitigation impact—intended to provide necessary third party due diligence information about different mitigation outcomes, in accordance with recognized rules and procedures".[17] As such, the NCM MV process tries to address an area of the MA's process that has not been elaborately addressed in the book. That is, once the quotas have been allocated and the CMO target has been established, how does an MA participant, with all the freedom to adopt its own carbon mitigation strategy, translate its national effort into the currency unit of the MA (i.e. a CMO) and how does the MA process guarantee that one country's mitigation effort is essentially, at the point of conversion into a CMO, equivalent to another country's effort based on a different mitigation approach model?

Besides the parallels, there are also overlaps between the MA proposal and the NCM work carried out by the World Bank to date. These overlaps include, among others, the idea of forming a club of countries interested linking towards a common system using the MV approach, conceptualising a specific body to act as the MV assessor for the countries participating within the NCM process and possibly, providing a country rating for its mitigation efforts. The concept of an MV assessor appears similar to the role envisaged by the book's author for the CMO-EB, whereas the idea that a country's mitigation efforts will be subject to a form of ratings system seems similar to a country rating to be issued by the MATF.

At the heart of the matter of the MV process, is a country accepting a third-party assessment and consequential publication of a qualitative and quantitative judgment

---

[14] See, the World Bank http://www.worldbank.org/en/topic/climatechange/brief/globally-networked-carbon-markets

[15] Macinante, J. (2016), Networking Carbon Markets— Key Elements of the Process. World Bank discussion paper, at p. 1.

[16] Macinante (2016), at p. 2.

[17] Macinante (2016), at p. 12.

regarding its mitigation policy and thereby being benchmarked against a notional ideal or those of other countries. The willingness to allow itself to be subjected to an assessment of this nature may be analogous to the economic review carried out by the IMF (under Article IV of the IMF Articles of Agreement) of a member country's economic performance through its country reports. Under the MA proposal, a country's willingness to subject itself to such scrutiny will be part of the MA membership requirements. In contrast, any 'pseudo rating' approach adopted may be analogous to that of a credit rating agency's Sovereign Ratings of a government's willingness and ability to service its debt on time and in full.

The NCM's biggest challenge is converting the conceptual model of MV into something that could work in practice. It is not going to be easy to establish a system whereby a country must accept that although their efforts quantitatively lead to the reduction of a tonne of CO_2 equivalent, qualitatively their efforts are unequal compared to the notional benchmark for effort or the efforts of a peer. There is, of course, a significant political sensitivity to such as assessment and no doubt a reluctance to accept any report that is not complimentary. Cooperation with the assessing agency and sharing significant of economic, environmental and fiscal data, which may be considered politically sensitive by many, will be key to the assessment process. This will raise concerns in many countries, surrounding the sacrifice of sovereign interests.

One of the differences between the NCM concept and the MA proposal appears to be the two different layers and levels at which the assessment process is to occur. The MA proposal potentially applies the qualitative assessment of a country's efforts at the quota allocation level and the quantitative assessment at the CMO-EB level. In contrast, the NCM approach is to have both assessments done by a single body. Although this difference doesn't change the nature of the concerns surrounding sovereignty issues, the nature of the body carrying out the assessment may be important to a country's willingness to submit to, cooperate with and finally accept an assessment. The MA proposals envisages that being carried out by an organisation with greater international gravitas than under the NCM proposal (which admittedly, is yet to consider this aspect more fully).

Ultimately, it seems that both the author of this book and the World Bank are seeking to address similar problems and are grappling with similar issues. However, both approach these issues from slightly different perspectives and it is informative to look to both in weighing up the various challenges and considerations.

The Complexity of Establishing the Governance and Legislative Framework

Chapter 8 of the book describes a proposed governance framework for the MA proposal. In so doing, it is inspired by the administrative and governance framework of the Bretton Woods Agreement and the institutions established thereunder. It is worth reminding ourselves that there is a distinction between a governance and legislative framework. Whereas the former can be considered the process "by which authority in conferred on rulers, by which they make the rules, and by which those rules and enforced and modified",[18] the latter is a broad system of rules that governs and regulates decision-making, as reflected in agreements, laws etc.

As such, the legal framework follows from the governance framework and it its end-product. The MA proposal is a broad inclusive proposal that, as conceptualised, is viewed as inclusive of national-level and sub-national level (e.g. regional governments) actors. However, the establishment of the bodies under the MA that have international recognition and status as public international organisations (such as the IMF, WTO or the World Bank), is likely to only be capable of establishment though international agreements that qualify as treaties within the meaning of the 1969 Vienna Convention on the Law of Treaties. Sub-national actors do not, under the laws of many jurisdictions, have the legal jurisdiction to enter treaties.

A practical limitation that therefore follows from the MA proposal to replicate the Bretton Woods governance framework, is that it then essentially becomes a government/country led process. The establishment of that process, particularly where intended to be consistent with a country's NDC (or within the MA process, the PDC), means that sub-national actors cannot lead in the process. They may be able to participate at a post-establishment stage, as may be the case for private sector actors, provided the MA framework allows it. However, that does not allow sub-national actors to therefore establish their own equivalent to the MA in circumstances where their country does not permit or support it. The MA proposal is therefore a more 'top-down' concept than the World Bank's NCM concept which seeks to ensure sub-national level participation through a less formal structure than existed for the Bretton Woods Agreement. This difference, partially, is the reason why I describe the MA proposal as ambitious.

To establish the MA proposal, to set up the applicable governance and legislative framework will require international negotiations, leading to the establishment of a treaty under international law that must additionally be consistent with or somehow recognised within the Paris Agreement. This almost invites the establishment of a protocol under the Paris Agreement (akin to the Kyoto Protocol to the UNFCCC). Although not all Parties to the Paris Agreement would need to participate in the negotiations or the establishment of such a MA Protocol, for the reasons outlined above, there will nonetheless will need to be a critical minimum number of participating Parties. The advantage of such an approach to establishing the framework of the MA would be the availability of a forum in which such negotiations could continue (e.g. on the side-lines of the CPA or UNFCCC meetings), the ability to support the negotiations with a secretariat that already existed (e.g. the UNFCCC secretariat in Bonn until the MA's own requisite body was established) and the assurance of the treatment of CMOs as consistent with the accounting and transparency framework that will be proposed for the Paris Agreement, in particular for ITMOs under A6PA, paragraphs (2)–(3).

[18] From the World Bank's website.

This is not to say that such negotiations would be obliged to follow the same painfully slow and often tedious negotiating formalities or processes of the UNFCCC but there is a risk of such issues being replicated if the negotiations were to be entirely consistent with the UNFCCC process. However, undertaking negotiations entirely outside of the UNFCCC process, with all the freedom to establishing new and more efficient negotiation approaches and procedures, also runs the risk of its eventual outcome (i.e. having reached an agreement) being held to ransom. Its eventual recognition, and more importantly of its administrative bodies, by the CPA under the Paris Agreement would then potentially be subject to the approval of, at least, the majority of the CPA Parties. At a minimum, such negotiations, should therefore engage the CPA and its secretariat during its negotiation process to facilitate its eventual approval and recognition.

Conclusion

Ultimately, the biggest difference between linking under the various proposals within 'carbon club' literature and the MA proposal is its ties to A6PA as a single cohesive set of approaches rather than separate and distinct approaches. The MA proposal, very ambitiously, seeks to solve all (if not most) of the Paris Agreement's failure to deal with the allocation of the Carbon Budget among its Parties by borrowing on elements of the approaches outlined throughout A6PA. In doing so, it is visionary and inspiring but creates for itself many significant hurdles. These hurdles are a mixture of political, financial, practical and legal.

Absent an environmental catastrophe or the visionary leadership among a coalition of the willing, it is difficult to see how the realities and practicalities of politics and economics will trump the instinct to protect national interests through protectionism and cross-border tariffs between those countries that are active in taking mitigating action on carbon and those that are not. The debate on carbon leakage within the EU emissions trading scheme has not nearly gained the intensity that is justified in a post-Paris Agreement context because the discussion has not been framed around a Carbon Budget. Once a Carbon Budget is recognised and accepted as a reality of the Below 2° Target, then the issue of cross border carbon leakage becomes more prominent and poignant. Through the MA proposal, this book's author envisages a solution for a problem that, in the minds of many, doesn't yet factor or register. His solution reflects the logical conclusion that can be drawn from, a deeply thoughtful insight on the impact of the Paris Agreement and the consequences that can follow from having a Carbon Budget. In that vision, he is very much ahead of his time. The book however, is an attempt to share that vision and invites discourse and debate, which from the author's perspective is only too welcome.

Glossary of Terms and Abbreviations

Accountability Situation in which someone is responsible for things that happen and can give a satisfactory reason for them.

Approach Exchange tool for mitigation outcomes and/or emission allowances.

Adaptation Initiatives and measures to reduce the vulnerability of natural and human systems against actual or expected climate change effects.

Additionality Reduction in emissions by sources or enhancement of removals by sinks that is additional to any that would occur in absence of the corresponding mitigating action.

Aggregate target Mitigation objective common to a group of stakeholders. It quantified what is termed 'common responsibilities'.

Allowance Emissions permit given to actors participating to cap-based emissions trading schemes based on a right-to-emit system (see permit).

Ambition Level of greenhouse mitigation voluntarily expressed by a party or stakeholder, with reference to a baseline status.

Article 6 of the Paris Agreement (A6PA) Article of the Paris Agreement adopted by this book as founding pillar for the proposed climate governance model.

Base year A specific year or an average over multiple years against which emissions are tracked over time.

Baseline State against which change is measured. It might be a 'current baseline', in which case it represents observable, present-day conditions. It might also be a 'future baseline', which is a projected future set of conditions excluding the driving factor of interest.

Baseline emissions The emissions that would occur without policy intervention.

Baseline methodologies Officially recognised methodologies for the implementation of emission's mitigation activities.

BASIC Group of countries, including Brazil, South Africa, India and China, formed to align negotiating positions on issues of common interest to these countries.

Border Carbon Adjustment (BCA) Policy measures based upon the application of import duties to safeguard the environmental integrity originated by domestic mitigation actions.

M. Stua, *From the Paris Agreement to a Low-Carbon Bretton Woods*,
DOI 10.1007/978-3-319-54699-5

Bottom-up Models that consider technological options or project specific climate change mitigation policies.

Burden sharing Distribution of determined mitigation responsibilities (quotas).

Cap-and-trade System establishing binding controls on the overall amount of emissions from participants. Within this quantity ceiling, entities covered by the scheme are free to choose where best to deliver emission reductions by trading units that correspond to quantities of abatement.

Capacity Combination of all the strengths, attributes, and resources available to an individual, community, society, or organization, which can be used to achieve established goals.

Carbon Abbreviation of Carbon Dioxide (see carbon dioxide) and adopted in this book as synonym of GHG (see greenhouse gases) thanks to the carbon dioxide equivalence.

Carbon budget Total amount of carbon permitted by climate security in a specific amount of time.

Carbon credit Generic term to assign a value to a reduction or offset of greenhouse gas emissions. A carbon credit is usually equivalent to one tonne of carbon dioxide equivalent.

Carbon dioxide (CO_2) Most abundant greenhouse gas in the atmosphere.

Carbon dioxide equivalent (CO_{2eq}) Metric measure used to compare the emissions from various greenhouse gases based upon their global warming potential. Carbon dioxide equivalents are commonly expressed as 'million metric tons of carbon dioxide equivalents' ($MMTCO_{2eq}$).

Carbon leakage Part of emissions reductions that may be offset by an increase of the emissions in non cap-constrained countries above their baseline levels.

Carbon market Popular (but misleading) term for a trading system through which countries may buy or sell units of greenhouse-gas emissions in an effort to meet their national limits on emissions (see also emissions trading scheme).

Carbon market mechanism Tool to facilitate the exchange of carbon credits.

Carbon offsetting Monetary investment in a project or activity elsewhere that abates greenhouse gas emissions or sequesters carbon dioxide from the atmosphere, which is used to compensate for greenhouse emissions of the investor.

Carbon Offsetting and Reduction Scheme for International Aviation (CORSIA) First global scheme aimed at neutralising the emissions effect of a whole industrial sector (civil aviation).

Carbon price Economic value placed on the emission of greenhouse gases into the atmosphere from human activity. A carbon price usually takes the form of either a carbon tax or as the cost of permits in an emissions trading scheme.

Carbon sequestration Process of removing carbon dioxide from the atmosphere and depositing it in a reservoir.

Carbon sinks Processes that remove more carbon dioxide from the atmosphere than they release. Both the terrestrial biosphere and oceans can act as carbon sinks.

Carbon tax Surcharge on the carbon dioxide equivalent content of actions envisaging production of greenhouse gases, aimed at discouraging their implementation.

Certified Mitigation Outcome (CMO) Certificate created within the Mitigation Alliance to certify the actual mitigation of carbon emissions. Issued through the Mitigation Alliance Mechanism, the CMO becomes the financial asset and representative currency for the whole Mitigation Alliance.

Certified Mitigation Outcome Bank (CMO Bank) Authority of the Mitigation Alliance (see Mitigation Alliance).

Certified Mitigation Outcome Exchange Board Authority of the Mitigation Alliance (see Mitigation Alliance).

Certified Emissions Reduction (CER) Reductions of greenhouse gases achieved by a clean development mechanism activity.

Clean Development Mechanism (CDM) Market mechanism adopted within the Kyoto Protocol.

Climate clubs Minilateral regimes having climate change mitigation and adaptation as priorities or side effects of the clubs' actions.

Co-benefits Benefits of policies implemented for various reasons at the same time. Most policies designed to address carbon emission's mitigation have other, often at least equally important, rationales.

Compliance Degree to which the actors whose behaviour is targeted by an agreement conform to the implementing obligations.

Conference of the Parties (COP) Supreme decision-making body of the United Nations Framework Convention on Climate Change.

Conference of the Parties serving as the meeting of the Parties to the Paris Agreement (CMA) Supreme body in charge to make decisions under the Paris Agreement.

Contraction and convergence Long-term strategy for managing global greenhouse emissions. Contraction refers to a declining global cap that would be set on worldwide emissions, together with a reduction trajectory over many decades. Typically, emissions entitlements would be allocated to nations by a formula that would converge over time towards equal per capita emission rights.

Domestic Offsetting Schemes (DOS) Recently designed and not yet implemented offsetting option that do not oversee the exchange of mitigation credits between stakeholders of different countries.

Double counting Distortive process that allows for a greenhouse mitigation to be banked by both its producer and the buyer of the corresponding carbon credit.

Emissions cap Mandated restraint in a scheduled timetable that puts a 'ceiling' on the total amount of anthropogenic greenhouse gas emissions that can be released into the atmosphere.

Emissions scenario Plausible representation of the future development of carbon emissions, based on a coherent and internally consistent set of assumptions about driving forces (such as demographic and socioeconomic development, technological change) and their key relationships.

Emissions trading scheme (ETS) Market-based approach to achieving environmental objectives. It allows those reducing greenhouse gases emissions below their emission cap to use or trade the excess reductions to offset emissions at another source inside or outside the country (see also carbon market).

Entry into force Point at which an agreement becomes binding. The entry into force of the Paris Agreement has been established at the point when at 55% of the signing parties, representing at least 55% of the global sources of greenhouse gases emissions, will ratify the Agreement. Ideally it will enter into force by the end of 2020.

Environmental integrity Safeguard of the biophysical processes that support nature and human beings continuity. In climate change it usually refers to the need for a greenhouse gases mitigation to be effective and additional.

Equity Fair correspondence between common but differentiated responsibilities and respective capabilities of the parties/stakeholders involved in a greenhouse gases mitigation process.

Excludable (Legal) opportunity to exclude specified actors from accessing to specified benefits.

Experimentalist governance (XG) Institutionalized transnational process of participatory and multilevel problem solving, in which particular problems (and the means of addressing them) are framed in an open-ended way, and subjected to periodic revision by various forms of peer review in light of locally generated knowledge.

Free riding Process to enjoy from the non-excludable benefits originated by efforts without participating to the efforts.

General Agreement on Tariffs and Trade (GATT) Treaty founding the World Trade Organisation and regulating international trade.

Global warming potential (GWP) Index, based upon radiative properties of well mixed greenhouse gases, measuring the radiative forcing of a unit mass of a given well mixed greenhouse gas in today's atmosphere integrated over a chosen time horizon, relative to that of carbon dioxide. The GWP represents the combined effect of the differing lengths of time that these gases remain in the atmosphere and their relative effectiveness in absorbing outgoing infrared radiation.

Governance Governing system that recognises the contributions of various levels of government (global, international, regional, national, local) and the roles of the private sector, of non-governmental actors and of civil society.

Greenhouse gases (GHG) Gaseous constituents of the atmosphere, both natural and anthropogenic, that absorb and re-emit infrared radiation. Greenhouse gases involved in global warming are carbon dioxide, methane, nitrous oxide (N2O), hydrofluorocarbons, perfluorocarbons and sulfurhexafluoride. Greenhouse gases emissions are calculated in carbon dioxide equivalent.

Group of 77 and China (G77+China) Group composed by more than 130 countries that act as a major negotiating bloc.

Human Development Index (HDI) Tool to assess countries' progress regarding social and economic development as a composite index of three indicators: (1) health measured by life expectancy at birth; (2) knowledge as measured by a combination of the adult literacy rate and the combined primary, secondary and tertiary school enrolment ratio; and (3) standard of living as gross domestic product per capita (in purchasing power parity).

Implementation Set of actions (legislation or regulations, judicial decrees, or other actions) that governments take to translate international accords into domestic law and policy.

Inclusive Offsetting Scheme (IOS) Offsetting system resulting from the Mitigation Alliance.

International Civil Aviation Organisation (ICAO) Private international organisation resembling the majority of international civil air carriers.

International regime Regime including any possible member referring to its domain.

Intergovernmental Panel on Climate Change (IPCC) International group mandated to publish scientific reports on climate change in the world and to evaluate the capacity of the environment and our societies to adapt and to propose measures.

Joint Implementation (JI) Mechanism to exchange mitigation outcomes introduced by the Kyoto Protocol.

Kyoto Protocol (KP) International agreement to reduce greenhouse gases, signed in 1997.

Land use, land use change and forestry (LULUCF) The total of arrangements, activities and inputs undertaken in a certain land cover type (a set of human actions). Land use change refers to a change in the use or management of land by humans, which may lead to a change in land cover. Forestry refers to the management of forests resources, especially as carbon sinks.

Monitoring, reporting, verification (MRV) Process/concept to support greater transparency in a climate change regime.

Members Determined Contributions (MDC) Evolution of the Nationally determined Contributions—NDC (see corresponding term) within the context of the Mitigation Alliance.

Minilateral regimes Regimes including a number of members of a domain that shall be higher than two but smaller than the totality of the domain's possible members.

Mitigation Anthropogenic-promoted action able to reduce or remove greenhouse gases and/or resources. In this book the term is used also as synonym for 'carbon emission's mitigation'.

Mitigation Alliance (MA) Climate governance model proposed by the book.

Mitigation Alliance General Assembly (MAGA) Authority of the Mitigation Alliance.

Mitigation Alliance Mechanism (MAM) Mechanism to certify and issue mitigation outcomes proposed by this book.

Mitigation Alliance Target Fund (MATF) Authority of the Mitigation Alliance.

Mitigation pathway Trajectory representing a cost-effective option to reach a specified greenhouse gases mitigation target in a defined timetable.

Monitoring, Reporting and Verification (MRV) Term describing all measures taken to collect data on emissions, mitigation actions and support, to compile information in reports and inventories, subject to some form of international review or analysis.

Montreal Protocol International treaty designed to protect the ozone layer by phasing out the production of substances that are responsible for ozone depletion.

Nationally determined contributions (NDC) Pledge-and-review approach adopted within the Paris Agreement to define the mitigation and adaptation targets of the participating parties. Before the entry into force of the Agreement the system is defined as intended nationally determined contributions (INDC).

Net zero carbon (NZC) Virtually null level or carbon emissions, reached by their reduction, avoidance, and/or sequestration.

Paris Agreement (PA) International agreement on climate change signed by almost 200 parties in Paris on 12 December 2015.

Party Entity that agrees to be bound by a treaty and for which the treaty has entered into force.

Peak year A year from which it is determined and agreed for the amount of emission (of a single entity or a group) to start reducing.

Permit Legal permission authorising the holder to emit a defined quantity of greenhouse gases (see allowance).

Per capita emissions Level of individual emissions for a country, geographic area or sector. It results from the equation that divides the total emissions of carbon dioxide equivalent of a selected country/area/sector by its population.

Pledge-and-review Bottom-up system to share of mitigation responsibilities between a group, based on voluntary pledges individually defined by the group's members, to be periodically reviewed and adjusted.

Plurilaterally determined contribution (PDC) Plurilateral approach to the implementation instruments of the Paris Agreement, as adopted by the Mitigation Alliance.

Private good Commodity or service providing excludable benefits to its owner.

Project design documents (PDD) Document required for the approval of Clean Development mechanism (CDM) projects. PDD must contain all information about the submitted project.

Public good Non-excludable commodity or service. Greenhouse gases mitigation is by definition a public good.

Quota Determined mitigation responsibility (see also burden sharing).

Reducing emissions from deforestation and forest degradation (REDD) Recognised system for mitigating emissions from land use, land use change, and forestry. REDD+ includes conservation and sustainable forestry policies as well as those aimed at preventing deforestation.

Scenario Plausible and often simplified description of how the future may develop based on a coherent and internally consistent set of assumptions about driving forces and key relationships.

Sequestration Process of increasing the carbon dioxide content of a carbon dioxide pool other than the atmosphere.

Share of proceeds Commission/fee established over the gains from the exchange of mitigation outcomes and used to cover administrative costs or actions in area other than mitigation.

Sinks Any process, activity or mechanism that removes a greenhouse gases from the atmosphere.

Social, Economic and Environmental Value (SEVMA) Value recognised to voluntary mitigation outcomes as previewed by Paragraph 108 of the Paris Agreement.

Stakeholder Person or organisation that has a legitimate interest in a project or entity, or would be affected by a particular action or policy.

Sustainable development Development that meets the cultural, social, political and economic needs of the present generation without compromising the ability of future generations to meet their own needs.

Target Reduction of a specific quantity of greenhouse gas emissions from a baseline date to be achieved by a set date or timetable.

Timetable Defined amount of time given to reach a defined target, or a share of it. In the context of this book it is synonym for timeframe, period and other time definitions.

Top-down Models evaluate the system from aggregate economic variables.

Transparency Quality of being done in an open way without secrets. In climate change transparency gains special significance as it relates to the accountability of the mitigation policies and actions, thus requiring safeguarding and control systems (see also measure, report, verification).

Uncertainty Expression of the degree to which a value is unknown.

United Nation Framework Convention on Climate Change (UNFCCC) International environmental treaty negotiated at the Earth Summit in Rio de Janeiro from 3 to 14 June 1992, then entered into force on 21 March 1994.

Value (economic) Benefit, usually expressed in monetary terms, gained from use or enjoyment from a good or service.

Voluntary Carbon Markets (VCM) Carbon markets whose functioning and structure is not based on binding emission caps, but is originated by the willingness of its participants.

World Trade Organisation (WTO) Intergovernmental organisation that regulates international trade.

FSC
www.fsc.org
MIX
Papier aus verantwortungsvollen Quellen
Paper from responsible sources
FSC® C105338

If you have any concerns about our products,
you can contact us on
ProductSafety@springernature.com

In case Publisher is established outside the EU,
the EU authorized representative is:
Springer Nature Customer Service Center GmbH
Europaplatz 3, 69115 Heidelberg, Germany

Printed by Libri Plureos GmbH
in Hamburg, Germany